Through Our
Eyes Only?

Through Our Eyes Only?

The search for animal consciousness

by **Marian Stamp Dawkins**

Mary Snow Fellow in Biological Sciences,
  Somerville College, Oxford
and University Research Lecturer,
  Department of Zoology, University of Oxford

Oxford • New York • Tokyo
OXFORD UNIVERSITY PRESS

Oxford University Press, Great Clarendon Street, Oxford OX2 6DP

Oxford New York

Athens Auckland Bangkok Bogota Bombay
Buenos Aires Calcutta Cape Town Dar es Salaam
Delhi Florence Hong Kong Istanbul Karachi
Kuala Lumpur Madras Madrid Melbourne
Mexico City Nairobi Paris Singapore
Taipei Tokyo Toronto Warsaw

and associated companies in
Berlin Ibadan

Oxford is a trade mark of Oxford University Press

Published in the United States by
Oxford University Press Inc., New York

First published by W. H. Freeman/Spektrum, 1993

First published in paperback by Oxford University Press, 1998

A catalogue record for this book is available from the British Library

Library of Congress Cataloging in Publication Data
Data available

ISBN 0 19 850320 2

Typeset by Keyword Publishing Services

Printed in Great Britain by Biddles Ltd, Guildford and King's Lynn *to Donald Griffin*

Contents

Preface

FOR AS LONG AS I CAN REMEMBER I HAVE BEEN FASCINATED by the question of what goes on in the minds of other animals. One of my earliest memories is that of sitting beside a pen of geese and wondering why they seemed so much less disturbed by a low-flying aircraft than I was myself, even though they had glanced up and evidently seen exactly the same object. I had then, as I have now, a sense of the mystery of having been born inside a particular skin and given access only to my own private experiences, cut off from direct experience of being inside other skins, other people's as well as those of geese.

This book is an attempt to give an account of what we now understand of the experiences of other species, without, I hope, losing the sense of mystery that any attempt to explain or even describe consciousness seems always to carry with it. It is aimed at anyone who has ever wondered about the phenomenon of conscious experience in themselves and the possibility of it in other species. It draws on some recent and significant discoveries in animal behaviour and shows how near – and how far – we still are from understanding what goes on inside non-human minds. It is a book which I hope will be read by people who are not scientists and yet would like to know what scientists have been up to in their quest for this, one of the greatest remaining of all biological mysteries. I also hope that some scientists will read it and get something out of it too. For none of us fully understands consciousness either in ourselves or in other species despite the great deal we have learnt about the bodies and behaviour of animals. But we have and are making progress and I want to show to as many people as possible how far we have got.

I have kept references to the scientific literature to a bare minimum in the text itself because I felt that to put in too many would be to run the risk of losing non-scientific readers, the very ones whose attention I was most anxious to keep. But I have supplied a fairly extensive reference list at the end, with an indication of which books or journal articles would be the best ones for following up particular subjects, so that anyone who wishes to can go further or check the details of what I have described. I hope this will be a reasonable compromise between readability and scientific respectability.

This has been an exciting book to write and I have been greatly helped by some extraordinarily helpful colleagues who took a great deal of time and trouble to criticize early drafts and make suggestions for improving the book, most of which I have taken up. Chief among these helpful critics are Donald Griffin, Aubrey Manning, Michael Hansell and David Wood-Gush. I am particularly grateful to Robert Seyfarth whose detailed comments steered me away from even worse blunders than inevitably remain. Mike Amphlett supplied some visual images that are exactly right and Gayle Stephens helped with the typing. In the background at all times has been Michael Rodgers, publisher extraordinary, who guided the book to completion with great tact and much appreciated support and enthusiasm through all its stages.

MARIAN STAMP DAWKINS

Somerville College, Oxford
June 1992

x

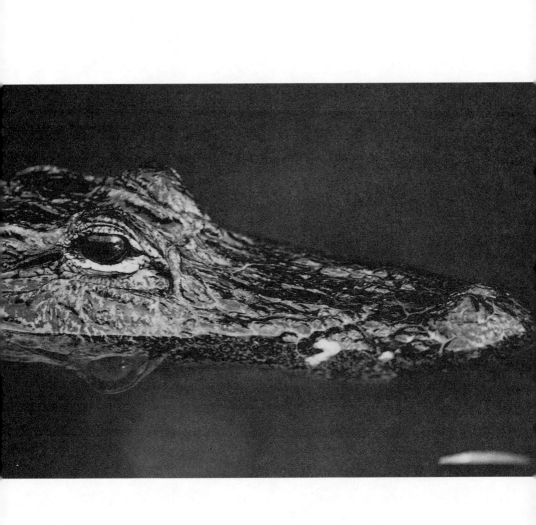

Chapter One

Through your eyes only?

IT IS AN EXTRAORDINARY FACT THAT EACH OF US SPENDS OUR ENTIRE LIVES locked inside the pinkish, blackish or brownish rubbery coating we call our skin. Inside this skin – apparently just behind our eyes – is an 'I'. 'I' is unique. No other 'I' can ever join us inside our skin or experience exactly what we experience. We are always on the inside and can know about other 'I's only from the outside of their skins.

But skins do not have to be prisons. What might seem like inevitable isolation from other people can be broken by experiencing a strong sensation that someone else understands what we are feeling and that we in turn can understand and share their feelings. We seem somehow to break out of the privacy of our skins and 'feel with' (the literal meaning of the word 'sympathy') other people.

What I want to do in this book is to explore the idea that some of the same processes of unlocking and understanding that we use with other people can be applied, at least to some extent, to other species of animals as well. Do other animals have conscious experiences at all like ours and, if they do, what are they like? Do they have thoughts and feelings? Are they aware of the world around them?

Such questions are not just difficult to answer; they call up all our reserves of ingenuity and scientific knowledge. Some people would even say that it is beyond the present state of science to answer them at all. Now, I would obviously not have written this book if I thought I was trying to do something that was completely impossible, but I do have to acknowledge how difficult it is to demonstrate the existence of conscious experiences in non-human animals and how much care is needed in evaluating the fragmentary evidence that we do have about what such experiences – if they have them – might be like. For this reason, I have written the book as if two very different groups of reader had to be convinced that the study of consciousness in animals is both worthwhile and feasible.

The first group is of people who, for various reasons, are sceptical about the idea that conscious experiences exist in any species except our own. Some, but by no means all these readers will be scientists who, because there is no 'proof' that other species have feelings, thoughts or emotions, refuse to take seriously the possibility that they might have them. It is as if they draw a ring around the human species and inside this ring they put themselves and other people. Outside the ring they put all of the rest of the animal kingdom, perhaps acknowledging that they may have complex behaviour and even have some striking similarities to ourselves in external appearance but refusing to believe that any of them are 'I's inside skins. If you are one of these sceptics or if you find echoes of this belief in yourself, I hope to convince you that the balance of evidence is now the other way. The evidence we now have, particularly from the study of animal behaviour, makes it simpler and more plausible to think that many other species do have conscious experiences than that they do not.

The second group of readers that I hope to reach take completely the opposite view. I know that for many people the idea that other species are *not* conscious is the ludicrous one. They are already convinced that at least monkeys, dogs and cats are fully conscious and that possibly many other species are as well. In fact, such people will probably see no need of a book to make a point they are convinced of already. If you take this view, I would still hope to persuade you to keep on reading and, paradoxically, to make you doubt what you already believe. For, despite what I am going to argue about the likelihood of conscious experiences in other species, I am also going to say that it is very difficult to know exactly what these are or even whether they exist at all. The assumption that animals are exactly like us except for having furry, feathery or scaly skins will emerge as just as ill-founded and just as misleading as the assumption that they are so unlike us that they could not possibly have any conscious experiences at all.

I shall thus be a sort of double devil's advocate, urging positive views on the sceptic and encouraging those who are already convinced to think again. This is not simply a game of picking holes in arguments just for the sake of it, but a genuine attempt to deal with the extreme complexity of the task in hand. Nothing will remove the fact that there are major obstacles to finding out whether other species are conscious. These obstacles, up to a certain point, can be overcome, but we have to be very careful in what we accept as good evidence for consciousness in non-humans because not all the evidence for it that is sometimes used is as good as it appears at first sight. Hence the appearance of a dance or game, seeming to flirt with one view and then turning round and finding fault with it. It is all part of the process of sifting and sorting, of examining the evidence and throwing it away if it is not good enough, to be left at the end with a much smaller but, I hope, more rigorously based pile of facts that will stand up to scrutiny and provide an awkward irritation to anyone who tries to maintain that only one species in the whole history of the earth has ever felt and experienced an inner, conscious life.

For it is the possibility that this inner life, this baffling, compelling sensation of being an 'I' inside a skin is not just ours alone but is shared with at least some other animals that will be the central concern of this book. Putting it this way throws up one of the many obstacles we shall encounter: the problem of defining what we are talking about in the first place. Defining consciousness is problematical not just because of the elusive, will-o-the-wisp nature of the phenomenon itself but because we can mean so many different things by the one word. We can be 'consciously aware' of a pain in the back or of a bird flying past the window – both immediate sensations and yet somehow the end result of a series of unconscious processes. Our eyes pick up the movement of an object, its colours and its approximate size. Quite unconsciously our visual pathways interpret this as 'pigeon' and it is this that finally pops into our consciousness. Then we can be conscious of images or memories not necessarily triggered by anything that has just happened to us. We can forget about someone for a long time and then suddenly bring the image of what they look like back into our conscious minds, images that must have been stored in some unconscious state all along. We can also consciously think about a problem and work out in our heads what we should do to achieve a desired end before we actually commit ourselves to doing anything at all. We might even have 'put off thinking about it' from yesterday, consigning the issue to some unconscious in-tray until we were ready to engage our conscious thoughts fully and decide what to do. Or, we may have strongly held beliefs about an issue that we will express and consciously think about when asked but which lie dormant, as it were, while the conscious mind tackles such immediate tasks as what to eat for breakfast or how to get out of the way of an approaching car.

Sometimes, of course, our conscious experiences reach rather higher planes, as when we wonder what consciousness is or whether other people mean the same thing by 'blue' as we do. On such occasions we are aware not just of the world but of our place in it. We have a sense of self separate from the rest of the world but possibly able to act on and change the world if necessary. We are self-aware and this for many people is the crucial part of human consciousness.

With so many different meanings of 'conscious' it is no wonder that it is difficult to produce a simple straightforward definition. Indeed, perhaps we should not even try to find one, for, given the evident diversity of conscious states, it could be positively misleading. I shall therefore use consciousness to mean simply an immediate awareness in any or all of the senses used above but stress that consciousness comes in many forms and that its nature is still deeply mysterious to us. To go any further at this stage would be to run the risk of what the American philosopher Daniel Dennett has happily called the 'heartbreak of premature definition'. We know a little and we do not understand fully. We can give a hint of a definition but not a complete one because for each of us there is a knowledge of what it is to be consciously aware and an inability to say what it is we know. It is as if a spotlight of attention plays across an empty stage and lights up the figures standing there. As it passes, the stage is transformed from darkness into light so that what was obscure is understood and what was hidden bursts into view.

It is this phenomenon, known vividly to each of us as individuals but impossible to define in any simple way, that we will be talking about. The question that will concern us is whether it is also known, at least in some measure, to other animals. This issue is of critical importance, because if we do decide that other species are conscious, then this will profoundly affect our attitudes and could completely change our ideas about what is morally acceptable to do to them. Eating them, doing experiments on them or killing them because they are doing something that inconveniences us could be seen in a completely new light. After all, the belief that other people are conscious, even though we can never find absolute proof that they are, is what guides many of our actions towards them. Where would morality be if we did not believe that what we did caused the states we call 'pain', 'happiness' or 'sorrow' in other conscious human beings? In other words, the assumption that other people are like us in having such experiences is basic to many of our ideas of what is 'right' or 'wrong'. We can believe that it is wrong to destroy an inanimate object like a rock or an expensive violin because these objects have some intrinsic value to us, but the wrong becomes a much greater one if what we are destroying is a human being and what we are doing is causing pain and suffering to a body with a conscious mind. Rocks and violins, however beautiful, lack feelings and human beings, however ugly, have them.

So one reason why the question of whether non-human animals have conscious experiences is important is because, if they do, we may be forced to bring them inside the circle that so many people draw around the human species. Different people will have different criteria for bringing them in or leaving them out. For some, evidence that animals can feel pain may be enough to extend some sort of moral concern to them. For others, the criteria may have to be stronger than this, perhaps not just that they can feel pain but also that they have an ability to be aware of pain and its causes. Yet others will only be impressed by animals that are clever and can 'think'. Some people may even demand evidence that animals are fully conscious in the same sense that humans are before they consider bringing them into the circle. The point is, however, that many peoples' ideas of morality, however diverse and however reluctantly extended to non-human species, centres ultimately on some aspect of what they perceive to be their 'minds' or absence of them, such as whether they can think or feel or are aware of what they are doing. So in order to decide how to treat other animals, most of us need to know something about what and how much other species experience. If we decide that other animals are not conscious, then possibly we can get on with our meals and eradicate pests and do all sorts of thing to them without being disturbed by the moral issues that might trouble us if we thought they were. Either way, we would do well to find out.

There is another reason, too, why it is important to know what other species of animal experience, a reason that has nothing at all to do with morality or deciding how to treat animals. This arises from the motives people have for studying biology in the first place. People study animals and plants because they want to understand the diversity of life and how all the organisms that inhabit the Earth evolved. They want to understand how their bodies work, how they reproduce and how a single cell can develop into a fully functional adult organism. And out of all the questions that still remain to be answered in biology, the deepest and most mystifying of all is that of consciousness: why is that we have this inner life of awareness, inaccessible to anyone else but of such importance to each one of us? Why did it evolve and how?

Scientists are just as mystified by these questions as anyone else, and although, as we will see later, they are now beginning to produce some tentative ideas, consciousness still remains an intractable and even embarrassing problem for biologists. It is embarrassing because it does not seem to fit into their usual evolutionary framework. It makes them feel uncomfortable, awkward, so much so that they will sometimes deny that it is a scientific problem at all, saying instead that consciousness cannot be studied scientifically and so cannot pose a problem for any scientific theory. But it is a problem and, what is more, it is a problem for one of the cornerstones of biology – Darwin's theory of natural selection.

As most people are aware, Darwin's theory is that the animals and plants we see today are the 'success stories', the survivors of various agents of destruction such as diseases or particularly clever predators that killed off their less well-protected contemporaries. The successful ones had some inherent protection, a superior immune system, for instance, or a skin colour that made them blend into the background that meant that they could survive where others, less well endowed, simply died. They then passed the keys to their success on to their children who grew up similarly benefiting from the physique and behaviour that had served their parents so well. This idea, which is still accepted more or less as Darwin proposed it over 130 years ago, has been immensely successful in explaining a wide variety of phenomena. But consciousness seems to be an exception, a major, intractable problem that stubbornly refuses to conform to Darwinian orthodoxy. Why should natural selection favour conscious organisms over unconscious ones? According to Darwin's theory, this must be because animals that are conscious are in some way better at getting through life and the business of reproducing

themselves than are unconscious organisms. But how does consciousness make animals better at doing anything? Every single, so-called
'function' of consciousness that has been proposed so far could just
as well be carried out – so it would seem – by an unconscious
organism or even by a machine programmed to behave in complex
ways. For example, if consciousness is thought to have something to
do with learning to avoid painful situations and the actual sensation
of feeling pain (hurting) is supposed to help this process, why couldn't
unconscious animals or machines learn just as well by simply avoiding
situations that had damaged them in the past? No conscious experience
of pain is strictly necessary – just a learning rule and a means of
detecting damage to the body.

This distinction is quite critical to what follows. Nobody would
question that it is advantageous for animals to have ways of avoiding
damage to themselves and keeping themselves out of harm's way.
But why do they have to be conscious to do it? After all, we
accomplish a great deal unconsciously – blinking our eyes if an object
comes too close, for instance, without consciously thinking that we
might be in danger until afterwards, or jerking our hands away from
hot stoves before we have consciously realized that our fingers might
be damaged. So effective mechanisms for avoiding damage to our
bodies are already in place. What does conscious experience add?
Why does pain have to hurt? Why can't our bodies get on with the
jobs of damage avoidance and damage limitation with a set of machine-
like rules without making everything so consciously *unpleasant* as
well?

In other words, there is still a problem. Natural selection
could not favour conscious animals over unconscious ones unless it
made a difference to them in some way that mattered and that we
could detect in their behaviour, perhaps in the speed or efficiency
with which they reacted to danger or anticipated the future. Here is
the potential embarrassment to the Darwinian theory of natural
selection. We don't know what that difference might be and, indeed,
some people claim that it would be impossible to detect that difference
even if it did exist. But if there is no difference between unconscious

and conscious organisms at how good they are at doing anything, then consciousness could not have evolved by natural selection because natural selection needs differences in efficiency of living to work at all. Either, therefore, conscious mental experiences did not evolve by natural selection and are different from all other aspects of animal life, a conclusion that would cause anyone who believed in the universality of Darwinism to shuffle their feet and feel distinctly uncomfortable. Or they did evolve by natural selection and they do make a difference to animals in some way but we have almost no idea how, an almost equally uncomfortable position to be in.

I shall argue later that this second view – that consciousness does make a difference to the way in which organisms possessing it function – is by far the most likely and that therefore the threat to Darwinism is more apparent than real. I shall also argue that this implies that consciousness can and should be studied by scientific methods and thought of as a biological phenomenon like any other, even though a very unusual one. But whether we see consciousness as so unique and mysterious that it will forever be beyond the reach of science, or as a temporarily baffling phenomenon that will one day be brought into science, conscious awareness is one of the biggest problems that our conscious minds can allow themselves to contemplate. At present, nobody understands it. Looking at the possibility that it may exist, in some form or another in species other than our own, is one way of trying to get to grips with one of the deepest mysteries in the whole of biology. If it seems that ours is the only species that has conscious experiences, then we would view this mystery very differently from the situation that would exist if we conclude that other species, in greater or lesser degree, also had similar stirrings of inner private lives. Our whole view of ourselves and the uniqueness of the human species could be changed by finding out whether consciousness is the sole prerogative of just our species or whether it is shared to any extent by other animals.

For two very powerful reasons, then, what non-human species experience is important to us. If they are conscious, this could change our view of how we should treat them. And it could also change our ideas of the evolution of our own consciousness. It is hard to think of any other subject that touches so deeply on so many important issues. Unfortunately it is also hard to think of anything else that is quite so difficult and intractable to study either. Its intractability comes from its essentially private nature. If you say you have a headache, I can hear your words and see the drawn look on your face. I might even, if I had the right instruments, measure various things about your brain, such as which parts were active and where pressure was building up. But having a headache is not just having symptoms that other people can detect. It is, of course, feeling the pain and being consciously aware of it. The private bit – what you experience inside your head – is the essential part of having a headache and it is also the bit that is inaccessible to me. You might, for all I know, be going through the motions of having a headache without experiencing anything like what I experience when I say I have a headache. I could probably guard against the possibility of your being a very good actor if I got out my instruments and examined you minutely and showed that the physiological symptoms you were showing were very similar indeed to mine when I have a headache, but I could never know for certain that head pains for you were the same as they are for me or even whether you consciously feel anything at all. It is for this reason that many people maintain that the study of consciousness is quite impossible. We can know what we experience on the inside of our own skins but our knowledge of experiences inside other bodies is strictly limited. The study of consciousness in other animals would therefore seem doomed to failure.

But there is a chink in this seemingly incontrovertible logic, a chink that we exploit every day in our dealings with other people and which we may be able to use for other species as well. Despite the impossibility of never really knowing what other people experience, we all of us go about our daily business as though we were perfectly well able to do so. We comfort babies. We write advertising slogans or say something that calms an angry person down. In a hundred thousand ways we all assume that we have access to what is supposed to be the inner worlds of others and the even more amazing thing is that the assumption seems to be valid. The child does stop crying and our angry colleague's temper is defused and he apologizes. By defying logic and acting as though we can know what other people are experiencing we can often gain control over a situation and predict what other people will do next. This strongly suggests that we are actually quite accurate in our assessments of what they are experiencing.

We manage to achieve this by a combination of two processes. Partly, we use our own experiences and assume that other peoples' are at least somewhat like ours. But we also do more than this and take into account the particular circumstances surrounding another person that might make them very different from us. For example, suppose you meet an old man whose cat has been killed by a car. For the sake of argument, let us say that you dislike cats and have none of your own. Your own direct experience and feelings about cats are therefore not going to be much help. The old man is not you, a cat-hater, losing a cat. Nor is he you, with your large family and busy life, losing a cat. He is someone whose own death is very close, who has no family living nearby and who has lost his only companion. Once you realize this, you can use your own feelings of loss, not of cats but of things that have mattered to you, to 'understand' someone whose life is very different from yours. You still use your own experiences – indeed it is impossible to get away from using them at some level if understanding is to be achieved – but experiences guided and modified by your knowledge of him and the sort of life he has been leading. You assume that he, too, has the inner awareness of grief and pain that you have known in another context.

Now, of course, an old man with a dead cat, even a taciturn, stoical old man who refuses to say very much, is altogether easier to understand than an organism of a completely different species. His appearance, gestures and tone of voice would say a great deal about what he is going through because he would be, despite differences in background and circumstances, like us in having a human shape, a human voice and expressions that move us because of their similarity to our own.

In principle, however, the same two processes that we use with apparent success with other people could be used with other species of animal as well. The 'same-but-different' worlds that we attempt to enter with other people become the 'not so similar and even more different' worlds of other species, but the means of entering them could have the same logic. Consider, first, the similarity to ourselves that gives us the sense that we can burst through the private nature of consciousness and know what other people are experiencing. We do this by what philosophers call the 'argument from analogy', by which is meant that we infer the nature of conscious experiences in other people by a leap of analogy with what we ourselves experience. We assume that because they are like us in ways in which we can observe (bodily shape and the possession of brains and faces like ours), they are also like us in ways that we cannot strictly observe, specifically in their possession of conscious experiences.

Our problem with other animals, then, is to decide whether other species are similar enough to ourselves that the argument from analogy – or some modified version of it – still holds. Can we enter their worlds of experience when those worlds are likely to be far more different from ours than other peoples' will ever be? One of the major themes of this book will be that there are far more similarities than many people realize and that the superficial differences, such as looking quite different or being covered with scales or feathers instead of bare skin, are far less important than the underlying similarities. If we look beyond the skin, beyond whether the animal lives in the sea or flies through the air, to the animal's own world, we find similarities that are at first not obvious at all. If we put aside the fact that, say, a particular animal carries out many of its social interactions through a system of sounds that are incomprehensible to us and concentrate instead on what the sounds tell it about fellow members of its species and on the insights it seems to have about what they will do next, again, we see analogies where previously we saw none. And if we allow that there are other ways besides speech that an animal can show that it has learnt something and understood the workings of its environment to a very high degree, then we have to pause for thought, for we are confronting a being that is 'like us' in ways that are uncannily like some of the ways that other humans are 'like us' too. Once we use the argument from analogy to break down the bulwark of the private nature of conscious experiences where other people are concerned, then the floodgates are open for other species too, particularly as we discover more about their behaviour and the complexity of their natural lives. The doors to even more distant worlds begin to open.

But this does not mean that, just because we have begun to see a way in which we might learn something about what other species experience, we automatically know everything there is to know about them, using nothing more than the intuition that serves us reasonably well with other people. Even with other people, there is a second process going on as we attempt to understand them – namely, the taking into account of the differences in life style and circumstances that exist between us. The headstart we get with other people because of their basic similarity to us has to be tempered, with more dissimilar organisms, with painstakingly acquired knowledge about them. This includes not just the obvious things about an animal, such as what it eats or whether it comes out during the night or during the day, but what its life is like. Does it, for instance, live in a world in which the threat of attack from a predator is so ever present that it cannot carry out any activity without constantly breaking it off to scan the horizon for danger? Is it a social animal, always in the presence of other members of its species or is most of its life spent in a solitary state? Does it rely on others to help it find food? The questions are numerous and we have to do a considerable amount of background work before we can accurately make allowances for the differences as well as the similarities between us and them. Otherwise, we will make the same sorts of mistakes we would have made had we assumed that the old man was relieved, as we might have been, that his cat had disappeared and that there was no longer the burden of having to look after it any more. Only this time, trying to cross a species barrier, our mistakes could potentially be much more serious.

13

Just as truly understanding another person can come about only if we take the trouble to find out things from their point of view and discover what matters to them, so, in the infinitely more difficult task of trying to understand another species, we can only succeed if we are willing to do the same thing, using what we know about its biology and the facts of its life style to curb our temptation to think they are exactly the same as we are. 'Anthropomorphism' – seeing animals as just like humans – can be just as erroneous as refusing to see any similarities at all. To go back to the parallel with our ways of understanding other people for a moment, if someone said to you 'I know exactly how you feel' when you had told them only very briefly what had happened to you, you would probably suspect that they didn't really understand you at all and were just projecting their own experiences onto you. But if they took the trouble to find out how you had responded to a particular set of circumstances and let you tell your story in your own way, then you might begin to feel that they had a fairly good insight into your state of mind. With other species, the assumption that a dog, a mongoose, a parrot or any other animal can be understood perfectly with no effort and no attempt to see things from the animal's own point of view will be even more likely to lead to a failure to know what they might be experiencing. Dogs are not humans any more than baboons are salmon or sticklebacks are reindeer. That is the reason why I said earlier that I hoped to spread a little caution among those who thought they already knew what other animals experience and have no need to study this further. None of us knows everything about such a difficult issue, and it takes humility to recognize that a human view of the world is not the only one there is. Other animals may have quite different experiences from ourselves but we are never going to discover what these are if we insist in squeezing them all into a human form. To be truly open to the discovery of what conscious experiences in other animals might be like, we have to be prepared to go beyond the narrow-minded, rather arrogant anthropomorphism that sees human conscious experiences as the only or even the ultimate way of experiencing the world and make ourselves open to the much more exciting prospect of discovering completely new realms of awareness.

The rest of this book is about the evidence we now have about what conscious experiences in other species may be like. It is written as a series of steps, each one building on what has gone before and each one making use of recent discoveries in animal behaviour. Non-human animals emerge as much cleverer and much more 'like us' than many people imagined in all sorts of ways, many of which are directly relevant to the issue of whether they are consciously aware of what they are doing.

Before we start to look at this evidence, however, there is one issue that we should get out of the way right at the outset because it constitutes for some people the one insuperable barrier to drawing analogies between humans and other species, the one way in which they are definitely not 'like us'. That barrier between them and us, that impassable moat, is language.

There is a very widespread view, one that I have encountered many times when I raise the question of consciousness in animals with a variety of people, that because non-human animals have no language as we understand it, they can have no way of telling us what they experience, even if they experience a great deal. An even more extreme version of this view is that without language, no thought is possible. So the absence of language not only makes it impossible for animals to tell us about what they are aware of it is also supposed to make it impossible for them to be aware of anything at all.

It is undeniably true that with other human beings, language in an extraordinarily important way in which we find out about their inner worlds. We use it to find out initially what someone might be feeling and later to refine and change our first impressions in the light of what they tell us. It appears, then, to be a major obstacle to our understanding of non-human animals that this channel is not open to us. Fortunately, however, while words are useful, they are not essential. We understand babies before they can speak and we do not need words to understand the suffering of famine victims whose language we do not know. For all the high-minded things we say about the uniqueness of human language and the way it sets us apart from other species, we often choose to ignore words in favour of what people do or even just a fleeting expression on their faces. 'Actions speak louder than words', we say, or, 'He put his money where his mouth is', giving more weight to what people do than to what they say. We have other means of finding out about what people are experiencing besides language and, as we shall see later, those same means can be applied to non-humans as well. In other words, the absence of language is a bit of a handicap, but it is by no means the insuperable difficulty it is sometimes made out to be. Without words, we just have to be a little more ingenious in the way we go about things, but we certainly do not have to give up altogether.

What we have to do to approach the important but elusive subject of possible conscious experiences in other species is to set aside the differences that in many ways do make our species unique and special and concentrate instead on what we might have in common, features that might – just might – lead us to the conclusion that other species too are conscious. This means that we have come right back to the central issue of whether any other species are sufficiently like us that we can safely use the argument from analogy, (which we use to infer consciousness in other people) with them too. As we have seen, it would have to be a somewhat altered version of the argument from analogy, tailored to take into account the fact that these other animals may not look at all like us or have anything like the same way of life, and that any conscious experiences that they might have might be totally different from anything we know about.

But if being sufficiently like us does not mean looking like us or having human customs and language, what does it mean? If consciousness does not have to have a human face, speak with a human voice or look out from human eyes, inside what sort of bodies should we look for it? And how will we recognize it when we find it?

Chapter Two

Miss Halsey
moves her foot

> . . . *if a beetle were to say to you, 'Please, Miss Halsey, will you avoid treading here, otherwise I shall be crushed', wouldn't you be willing to move your foot a trifle?*
>
> Fred Hoyle, 'The Black Cloud'

IMAGINE FOR A MOMENT THE SITUATION THAT WOULD EXIST IF THE EARTH were suddenly invaded by a group of large green blob-like creatures that took us all over, used their evident technical superiority to subvert our machines, spoke to us in our own native languages and treated us with apparent concern. We would all, of course, be confronted with major problems, not just of how to get on with our lives, but of how to treat the green blobs themselves. We would probably be angry and frightened, but we might also have to admit, if we found time to think about it amidst the chaos, that despite their very unhuman appearance and disconcerting habit of neither eating nor defecating, they had some things in common with us. If we found that they had what seemed to be courts of justice that tried offenders and that, further, they seemed to look after their strange globular offspring with enormous tenderness, wouldn't we eventually have to put aside our deep-seated prejudices that green blobs could never be said to be remotely 'like us' and at least consider the possibility that they might have conscious experiences going inside their slimy, green, globular skins?

It would, of course, be possible that all they were doing was 'going through the motions' without any accompanying conscious states – caring for their offspring, for example, with no real parental feelings to accompany the movements of protection and going about their business with no conscious thought about what they were doing. But if their behaviour were sufficiently complex, then we would surely begin to conclude that it was much more likely that there were minds at work behind the behaviour – thinking, feeling minds, that is, with private experiences at least in some ways like our own.

So one of the first criteria that might make us think twice about consciousness in organisms that look very different from us is the complexity of their behaviour. This is not to say that all complex behaviour implies consciousness but, rather, that the complexity of behaviour and the ability to adapt to changed circumstances are some of the hallmarks of a conscious mind. Complexity could, of course, be achieved without consciousness, as computers with their fine control over missiles or car assembly plants will constantly remind us. But the more we find that an organism is not just following set routines and is able to adapt its behaviour to overcome obstacles put in its way, the more plausible it becomes that it is achieving this through conscious thought. The more similarities we can see between what we do and what they do, the more plausible it becomes to infer that they are 'like us' in having conscious experiences to achieve that complexity. If we can grant this possibility to hypothetical green blobs, how much more reasonable it is to grant the same to the real animals that are all around us, share a common evolutionary heritage with us and have nervous systems made from the same sorts of cells.

The first part of our quest for consciousness in other species, then, is to look at what we now know about their behaviour and to see whether this can be dismissed as simply the workings of unthinking, unfeeling automata, blindly going through a set series of actions with no hint of a mind to guide them or whether what they do is complex enough and unpredictable enough that it suggests the possibility – even a remote one – that somewhere a spark of consciousness has come alight.

A stumbling block that many people have when they look at animals is that they see their behaviour as being the opposite of clever and complex. In fact, they see animals as essentially stupid. They will use examples like a bird fighting its own reflection in a mirror and say that this shows that the bird is very unintelligent because it apparently never realizes that there is not a real rival there. Or they point to a dog turning round and round on a carpet before it lies down and say that this, too, is stupid because the dog does not realize that the grass it is attempting to flatten is non-existent. Their implication is, of course, that humans could never behave in such mindless, unthinking ways. They conveniently forget cases where they appear to, such as when a man responds sexually to the photograph or model of a woman as if it were one. I have had it confidently stated to me by an eminent philosopher that humans alone have the power to think about their actions, to learn from their mistakes and to choose how to behave, whereas other animals do not. Non-humans, according to her, are governed by their instincts and blindly follow what those instincts tell them to do.

Now it is indeed true that some of the behaviour shown by animals is what can be called instinctive or 'innate' (most usefully and uncontroversially defined as 'unlearnt') and that this does sometimes lead them to pick out a particular feature of their environment and respond to it apparently stupidly as though it were the real thing. Niko Tinbergen, one of the pioneers in the study of animal behaviour, used to tell a famous story of how some sticklebacks he was keeping in his laboratory at the University of Leiden once became extremely aggressive when they saw a red mail van passing outside the window where their tanks were kept. Sticklebacks are small fish in which the males develops bright red coloration on the undersides of their bodies during the breeding season. The passing van bore only a remote resemblance to a male rival of their own species, but its red colour was enough to set off the behaviour they would normally show to a real fish – namely, attempting to fight it and chase it off their territory.

But just because it is possible to pick out a few quaint and amusing examples of animals behaving in simple ways and not seeming to realize their mistakes, this does not mean that their lives are wholly made up of blind, innate responses to the world around them. On the contrary, the more we learn about animal behaviour, the more it seems that such stupidly simple responses are the exception and that a far more complex assessment of the animal's environment is the rule. In fact, there are numerous cases where the animals remain several jumps ahead of the humans who are investigating them. Nothing better illustrates this than the case of vervet grunts.

Vervets are small, slender monkeys with black faces and long, graceful tails that they hold in an arch behind them. They can climb trees with ease, but spend a lot of their time looking for food on the ground. They are above everything very social animals, staying all the time in the same small group, grooming each other, feeding together and warning each other about possible dangers. Their system of communication is immensely complex and they have three different alarm calls for different sorts of predator – one call for 'snake', another for 'leopard' and yet another for 'martial eagle'. But it is not just these 'danger' calls that give them what might be called a rudimentary 'vocabulary' and forestall any description of them as simple or stupid. It is their extraordinary collection of grunts as well.

Vervet grunts are soft, repeated noises which are low-pitched, short and sometimes described as a series of 'woofs'. Human observers had heard these grunts for many years and concluded that these were just some sort of vague way the monkeys had of keeping in contact with each other. Then, Dorothy Cheney and Robert Seyfarth, who had been making a long-term study of these monkeys in the Amboseli National Park in Kenya, decided to record these grunts and analyse their composition. They used a sound spectrograph, which is a machine that analyses the frequency of sounds and gives a printed record of the different pitches throughout its length. To their surprise, they found that the machine did not give the same sound picture for all the grunts. It showed four different patterns of grunt, even though to the human ear they all sounded the same. Cheney and Seyfarth realized that these different grunts were given in different circumstances, one type being given when a monkey met another monkey that was socially dominant to it, another when it met a social inferior, a third when it was moving into an open space and yet another when the monkey saw a group of strange monkeys. The exciting part came when they played back recordings of these different sorts of grunt to a troop of wild monkeys. They hid a tape recorder in vegetation so what a monkey would hear would be as similar as possible to its hearing a sound from another monkey temporarily out of its sight. They also filmed everything that happened before, during and after each recording was played so that they had a permanent record they could look at in detail afterwards. Choosing their moments carefully, they switched on the tape recorder near an unsuspecting vervet. The monkeys behaved exactly as though another monkey was grunting in the bushes. If the grunt was of the type normally given when moving into an open space, the monkey hearing the tape recording would look not at the loudspeaker, but into the distance, as if anticipating that the whole troop was about to move into that area. If the monkey heard from the tape recorder a grunt usually given by a dominant to a subordinate animal, then it would tend either to stare hard in the direction of the speaker or move away from it, whereas if it heard the sound of a subordinate meeting a dominant animal, there was much less response, suggesting that the monkeys could definitely tell the difference. On hearing the grunt usually associated with seeing monkeys from a strange troop, the monkey would look quickly at the speaker and then in the direction in which the speaker was pointed, apparently alerted to the possible presence of strangers in the area.

So the monkeys are able to pick up subtleties in their grunts that completely escape the human ear. To a human, a grunt is a grunt is a grunt. It takes technological aids in the form of tape recorders and a sound spectrograph to show that, a far as the monkeys are concerned, there is far more to it than that. Slightly different grunts evidently convey quite different messages to the monkeys that hear them – the very opposite of a crude, simple response to a stimulus. We still do not fully understand what it is that the monkeys are responding to or exactly how they manage to detect the differences between the grunts. But do it they clearly do, and they leave their human observers slightly baffled.

In some ways, the feats performed by female ostriches are even more remarkable and leave us even more mystified as to how the trick is done. By any criteria, ostriches are bizarre birds, not just in their ungainly appearance but even more so in their reproductive habits. Their system of child-rearing defies any normal description. Pairs of ostriches often kidnap each other's chicks and form huge mixed broods with some of their own and some of forcefully adopted chicks mixed together. The successful kidnappers will chase off the real parents and look after all the chicks as if they were their own. However, their entire haul of chicks may in turn be stolen by yet another pair of ostriches, who will make an ever bigger brood by adding their own to the newly stolen ones. This strange, serial kidnapping appears to be a way the parents have of protecting their chicks through a form of dilution. Any flock of ostrich chicks is going to be very conspicuous to predators anyway, so having some extra ones around that are not theirs increases the chances that when a predator does strike, the chicks taken will not all be the parents' own. They effectively hide their chicks in a living shield of other parents' chicks and the more of these other chicks they can manage to accumulate, the better protected their own chicks will be. Hence the frantic race to capture as many other chicks as possible. But apart from their forcible baby-sitting, ostriches do something even more extraordinary with their eggs.

Up to half a dozen female ostriches will lay their eggs in the same crude nest, which is really nothing more than a shallow scrape in the ground. A single nest may eventually have as many as 40 eggs in it, each one weighing 1.5 kilograms but they will be incubated and guarded by just one female, known as the major hen, helped by the male. The major female, however, can manage to incubate only about 20 of the eggs in her nest and she pushes the surplus eggs out of the nest where they perish. But she does not put out eggs at random. She has laid many, but by no means all the eggs in the nest and she sees to it that it is her own eggs that are incubated and those laid by other females that are mostly pushed to the edge. She apparently recognizes which eggs are her own and which were laid by other females, and gives her own eggs top priority. Brian Bertram, later Director of Research at the Wildfowl Trust in Slimbridge but then working on wild ostriches in Kenya, carefully numbered each egg in various nests as they were laid, so that he knew which eggs were laid by which females. He found that the ability of some females to tell the difference between their own and other birds' eggs was absolutely perfect – they never pushed out their own eggs and always moved the eggs of other females to the edge of the nest. A female ostrich did not simply use the age of eggs as a guide to which ones were hers, because she was equally likely to throw out old as new eggs if they did not belong to her. Nor did she use the eggs' position in the nest because if Bertram moved the eggs around to new places where the female had not laid them herself, she still seemed to know which were her own eggs. He was forced to the unlikely conclusion that the females were recognizing their eggs by the pattern of pits on the surface. All eggs have tiny 'air holes' in the shell to allow the developing chicks to breathe. Differert eggs have slightly different patterns of holes, so slight that it is difficult to believe that a bird could possibly notice the differences. But it seems that this is exactly what the females are doing. Like vervet monkeys, they are making discriminations that humans find it nearly impossible to believe that they can make, let alone make themselves.

So it is already clear that not all animals make simple responses to crude stimuli. Baffling complexity rather than simplicity seems to be the best description. And the more examples we come across, the more this seems to be true. In those two major areas of any animal's life, sex and aggression, for instance, we find not only that animals can make complex discriminations about each other when they encounter them but that they even have an uncanny way of predicting what they might do in the future.

When an animal is confronted with another of its species that challenges it to a fight, it is faced with a dilemma. If it fights, it stands to gain, or at least not to lose, something of value to it, such as food, territory or a mate. On the other hand, it may be so severely injured during the fight that it will be killed or at least put out of action for the foreseeable future. But if it does not fight, while it may be safer, it also runs the risk that it will lose the very things that are the keys to its survival and successful reproduction: mating opportunities, perhaps young that have already been born and often life-saving food supplies. Now, the outcomes of fights are not completely random, even though they may have an element of luck in them. Just as the outcomes of human boxing matches are to a certain extent predictable from knowledge of past form and the relative size and strength of the two participants, so, too, are the outcomes of animal fights both to us and, it would seem, to the animals involved as well. Sometimes these predictions would be so obvious to be uninteresting. Very small animals tend not to get involved in fights with very large ones – the odds would be too much against them. In the same way, pitting a heavyweight boxer against a flyweight would be pretty much of a non-contest and if the flyweight had any sense he would (like the very much smaller animal) refuse to fight in the first place. That is why there are so many weight classes in boxing. The people who do end up fighting each other each have a reasonable chance of winning because they are of roughly the same physique and ability. This does not stop speculation about who might actually win, based on slight differences in weight, health or physical condition on the night but essentially boxing matches are staged between fighters sufficiently similar to each other that the outcome of their match is in some sort of doubt. It is, of course, the element of unpredictability that gives the point to the fight and makes it worth watching.

Now, animals do not have ready to hand the detailed information about their opponents that a human might have, such as how he performed in a fight last week or exactly at what level he tipped the scales that evening. But they do seem to be able to make an extraordinarily accurate, on-the-spot assessment of their opponent's fighting ability and the likelihood of his winning a fight. The risks and potential benefits of fighting are so finely balanced that any information an animal can gather about how likely he is to emerge victorious is going to be vitally important in his decision of whether or not to go ahead and fight. (The question of whether this 'decision' is conscious will be left until later.) If it looks as though the opponent is stronger, fitter or simply more capable of sustaining a long fight, then the best strategy might be to withdraw without a fight. But if it looks as though the opponent is exhausted, or not very well muscled, then the decision to attack might have a much more profitable outcome. There will always be an element of chance – for instance, a stronger opponent whose foot slips at the wrong moment may lose – so even a formidable rival, provided he is not so much stronger as to be the inevitable winner, will sometimes be worth challenging, particularly where the stakes are high and the winner takes all.

If this all sounds fanciful and a bad case of the anthropomorphism that I have already argued can be so misleading, let me assure you that many animals, even without the benefit of form books and statistics from ring-side commentators, do seem to have a very precise knowledge about the probability that they will win a fight against a given opponent. They are able to weigh each other up, assess another animal's fighting ability relative to their own and make what appears to be an entirely rational decision to fight or withdraw on the basis of their assessment. They show the very opposite of the impulsive, simple response to a challenge from another member of their species that we would expect if they were behaving 'stupidly' or just picking out a few crude elements of their environment to respond to. They certainly do not fight every time a simple stimulus is presented to them. On the contrary, as we have seen before, it is often the human observers who are at first puzzled as to how the animals are making the complex discriminations that they undoubtedly are and who have to make quite detailed investigations themselves before they understand what the animals are doing. This time our example comes from the fearsome mating battles of red deer stags.

On the remote island of Rhum, which lies off the west coast of Scotland somewhat south of Skye, Fiona Guinness has for many years been studying the red deer that live there. In conjunction with a team of researchers from the University of Cambridge led by Tim Clutton-Brock, she has watched and observed the deer so that she now knows all the animals as individuals and knows which are their fathers and mothers, grandparents and grandchildren. Because the deer are confined on the island, there is no possibility of their mixing with other deer from the mainland, so she can accurately record the pedigrees of each one and the history of their births, matings and death.

For most of the year, the deer graze peacefully, the females keeping together in loosely tied groups, and the males more spread out or in small groups. But in the autumn, the peace is shattered. The larger, stronger males become extremely belligerent and round up groups of females which they vigorously defend against other males. The smaller males, usually the younger ones not yet in possession of a group of females, spend their time either challenging the larger males or mating surreptitiously with any stray females they can persuade to co-operate. The air is filled with the raucous, gutteral sound of the stags bellowing at each other and the occasional clash of real battle. This is the time of year when the females are briefly receptive to mating and when all the young for the following year are conceived. Because the females are in groups anyway, it is possible for a strong male to round them up and keep for himself a large number of matings, provided, that is, he can ward off all the other males that have no mates at all and also have their eyes on his reproductive prize. One male can have 20 or more females in the group he is defending, but whether he manages to sire their offspring will depend entirely on how vigilant he is in seeing that they do not stray and how effective he is in fighting off the constant challenges from other males. His labours last for a long time. For 5–6 weeks he must be on constant guard, fighting and watching. Often, he has little time to eat, so his health may deteriorate, whereas his challengers, not having females of their own to defend, may become relatively stronger. This means that a stag that starts off the mating season able to beat all-comers may end it having to struggle against his younger opponents and even, eventually, lose the females that he has been so diligently defending. The outcomes of fights, even between the same two stags, is in a constant state of flux as the mating season takes its toll.

All through the autumn, from mid-September to the end of October, the battles over matings rights continue. Some of the stags are severely injured, losing an eye or suffering wounds to their neck or flanks from the sharp antlers of their opponents. Clutton-Brock estimated that each year about 23 per cent of the stags incur injuries severe enough to disable them or at least jeopardize their chances of surviving the coming winter. But he noticed that, as well as the serious fights, there were also a lot of what appeared to be non-fights, encounters between stags that came to nothing, a challenger retreating without having come to blows at all. All that happened was that the stags spent considerable amounts of time roaring at each other, and then broke off their encounter. But the roaring did not seem to be completely unrelated to real fighting. Fights tended to occur after an extended bout of roaring, with first one stag and then the other delivering a burst of roars or bellows with considerable force. Only after one of these 'roaring matches' had gone on for some time and the stags had spent several minutes pacing up and down looking at each other would they put their heads down and engage their antlers. Together with Steve Albon, Clutton-Brock set out to test the idea that roaring matches were not just 'threat' as most people had supposed up until then, but were the stags' way of trying to gauge each other's strength and so to work out which one was likely to win a fight before it actually happened.

It was immediately obvious that the stags had at least a crude idea of who was likely to win because, like boxers sorting themselves into weight classes, they tended to fight only opponents that were roughly their match in weight, size and known dominance rank. A young stag, attempting to challenge an older, larger and socially dominant stag for possession of a group of females, would approach, perhaps roar a few times, receive a few roars in reply and then retreat. To a human, as apparently to the deer themselves, the contest was so uneven and the outcome of the fight so predictable, that it was not worth pursuing the challenge. But if the challenger was more comparable in physique, then the outcome could be quite different even though a fight might not occur for some time. The hopeful challenger would approach and roar. The defending stag would roar back. The challenger would reply but this time he would deliver his roars at a higher rate, putting more effort into his challenge because roars are exhausting to produce and the stags appear to convulse their chest muscles each time they roar. The defender would then reply but even more forcibly and at an even higher rate, escalating the contest and apparently goading the challenger to roar even faster if he could. The challenger would then either go one better and roar at a yet higher rate or withdraw altogether. Roaring matches could continue until both stags were utterly exhausted by their efforts, to the point where neither would, for the moment, be capable of sustaining a long fight until they had got their breath back. They would then often walk up and down, looking at each other sideways on, examining the shoulder muscles that are used in fighting. Only if this examination failed to reveal any major differences would a real fight take place.

Clutton-Brock and Albon showed that the roaring matches and inspections gave the stags information about each other's physical prowess and so enabled challengers to weigh up the likelihood that they would win any potential fight. Their 'predictions' about the future are made from the roars because of what roars say about the stag that gives them. Producing a whole series of roars in quick succession takes some doing and can in fact be achieved only by a stag that is in good physical condition; in other words, one that is also going to be able to sustain a prolonged fight because both roaring and fighting depend on a stag having strong chest muscles. Ability to roar at a high rate will therefore be closely correlated with ability to win fights. Roaring can thus be used as a reliable – and safe – indicator to a challenger as to the likely outcome of a fight should he decide to attack. A stag that can out-roar him, producing many loud roars in a short space of time, will almost certainly out-fight him and should be avoided. But if the war of the roars and the detailed inspection of the opponent's body fail to separate the two contestants, then there is nothing for it but to put it to the ultimate physical test.

The fact that the stags do fight under circumstances when no predictor of fight outcome seems possible and withdraw from encounters when some critical difference between them becomes apparent strongly suggests that they are indeed using detailed information about each other to assess their chances of winning or losing. If the difference is not immediately apparent, as in some obvious discrepancy in body size, then they start probing each other, demanding proof of physical stamina and prowess in the form an increasingly exhausting regime of roaring, which only the genuinely physically fit can cope with. This is, of course, a far more reliable method of gauging an opponent's true worth on a given day than a more obvious trait such as antler size. When a stag grows his antlers, he may be a fine, healthy specimen. But during the course of the breeding season, when he fails to eat enough because of all the fighting he has been doing and may lose up to 20 per cent of his summer body weight, his antlers will remain the same size, but his fighting ability may decline dramatically. And what his young opponent wants to know is not how good a fighter he was at the beginning of the season but how good he is now, today. His antlers may not give him away, but his ability to roar will. Disregarding the large body size and the imposing antlers, but sensitive to the fact that the older stag has roared at a rate of only six roars a minute but he himself has achieved eight roars, he looks at the slightly shrunken shoulder muscles, paces up and down for a while and then goes in for the attack. The females have a new, younger mate as the father of their calves next year.

The decision of the younger stag to attack, however, is not one that was taken quickly or easily and it is certainly not a case of his 'stupidly' picking out one simple feature of his opponent and responding just to that. On the contrary, a great deal of time is taken to weigh up an opponent and the subtlest cues are looked for in case they reveal a critical weakness. If the cues are not immediately available, they are elicited, so that a weakened opponent cannot help but betray himself. What the animal eventually does is to respond not to simple cues but to those that give the best possible indication of what might happen. People placing bets on the outcome of boxing matches or horse races or elections rarely seem to do much better.

Similarly subtle weighing up of cues is often shown by female animals when they choose a mate. It would be easy to think that all that matters to a female is that she chooses a male of the right species, but the more we discover about female choice, the more it seems that females are incredibly fussy about who they mate with. It is not enough that he is the right species or the right age or ready to mate with her. He must have other attributes, and sometimes exactly what those are may be obscure to the point of complete mystery. An example we can use here, because it really does show how complex female choice can be and because it has genuinely baffled the people trying to unravel it, is the 'life prospects' policy of the female black grouse.

Male black grouse are gorgeous birds. They are powerfully built and have glossy black plumage, bright red combs and wattles, and a stunning white 'lyre' tail that juts out on either side. The tail is spread and fanned during the courtship display and if the female was going to respond in a simple way to the male, she would surely be seduced by this large white lyre spread out against the black of the rest of the feathers. Females certainly have plenty of opportunity for being stimulated by this obvious ornament. All the males in a given area gather in one place, each one defending a small bit of ground around himself and each one displaying to any passing female. These communal display grounds are called 'leks' and are the bird world's equivalent of a marriage market. The female black grouse come to the leks and walk around them, examining the males before they eventually decide which one to mate with. Jacob Höglund, who with his Swedish colleagues, Rauno Alatalo and Arne Lundberg, has made a long-term study of these birds, believes that the females walk down the boundary between the territories of two males and provoke them to fight, thus giving themselves a direct indication of which one is the stronger by seeing who wins. Whatever criteria the females are using, they eventually choose one of the males and mate with him. The obvious thing for the females to do is to choose the males with the largest white tail, the biggest body size, the reddest wattles or some other clearly defined feature. But the female black grouse seem to have other ideas. They certainly mate with some males more than with others but Högland and his colleagues were completely unable to work out what they were using as the basis of their choice. The favoured males did not seem to be the ones that were necessarily larger or displayed most vigorously. Nor were they the ones that obviously appeared to be the healthiest, at least as judged by the number of parasites they were carrying. True, a male in demand among the females often seemed to have a large white tail but then so did a lot of the males that were inspected and subsequently ignored by them. A beautiful tail seemed to be a help in attracting females but not the whole secret.

After measuring everything they could think of about the males that were chosen and the ones that were unsuccessful in getting matings – size, weight, tail, comb, wattles, display, position in the lek – in an attempt to work out what it was that the females were going for, the Swedish workers were about to give up and say that there was no real pattern in the choice that they could detect, when they noticed something very strange indeed. Although their many detailed measurements gave them little clue as to what criteria the females were using at the time when they made their choices, the females seemed to be unerringly picking out the males that would be still alive in 6 months time, even though the females had nothing more to do with the males after they had mated with them and so could not possibly be helping them to stay alive. In other words, it was impossible for the researchers to predict at the time of mating which males the females would go for on the basis of appearance or behaviour, but if they waited for 6 months and then recorded which of the males were alive and which had died in the meantime, they could see that the females had 'predicted' the males that would survive and had chosen them to be the fathers of their children.

Högland and his colleagues themselves were unable to make this prediction. To them it seemed a matter of chance which males would survive and which would not because the obvious signs of health or impending sickness, such as how many parasites a male was carrying or how much body fat he had, had surprisingly little to do with his prospects of surviving the next half year. A male that had a small number of parasites and a healthy body weight during the mating season might subsequently die for some reason, while another whose condition seemed much the same or even slightly worse could still be alive and well 6 months later. But the female black grouse had evidently picked up some subtle cue that had escaped the investigators. While inspecting the males or watching their fights or generally seeing how the males interacted with each other, they must have detected some slight weakness in some males that was a sign of trouble ahead. These doomed males they ignored in favour of others that, in ways that we still do not understand, differed and had a much brighter future. As far as their chicks were concerned, mating with such males was probably the best thing the females could do because whatever it was that made the fathers likely to survive could well be carried over into the next generation and make the chicks more likely to survive well too. We still do not understand how female black grouse perform what amounts to a prediction of the future but it is quite clear that they are far less impressed with the display and showy feathers that are paraded in front of them than with some much more subtle cue that will in the end probably have considerably more bearing on the survival chances of their offspring.

We have seen enough now to realize that much animal behaviour is far from simple. There are a few cases where animals can be seen to pick out just one or two features of their environments to respond to them and where they can be easily fooled by a relatively crude representation or model, like the sticklebacks with their mail van. But these are not, as we have seen, general illustrations of the way all animals behave. The vervet monkey calls that need sound-analysing equipment to sort out differences that the monkeys themselves use all the time and the ostrich hen performing the near-impossible task of sorting eggs laid by herself from those laid by other females both show that the discriminations that animals make can be extremely complex, much more so than we may give them credit for. It would have been easy to dismiss non-human animals as simple and to have overlooked the complexity of their responses to the world had these studies not been done. Fortunately, more and more such studies are now being reported and what many of them have shown is evidence of animals behaving even more 'cleverly' than most people imagined. Furthermore, animals do not just give an immediate response that could be described as complex, but may also go so far as to search for and elicit extra information from their environments before they decide what to do next. The probing examinations that red deer stags give each other before committing themselves to a fight and the detailed inspections that female black grouse make of males before they mate with them both show how animals may take a great deal of time and trouble to gain information that may not at first be available to them. What they are doing is the very opposite of behaving 'stupidly' and blindly. On the contrary, they seem to be being extremely clever indeed.

But are they really? Isn't it possible that all this apparently clever behaviour is the result of some very simple pre-programmed responses that show no evidence at all for any real understanding? Perhaps the female black grouse are responding to some minor differences between the males – a slight difference in wattle colour, for instance – that humans have not yet detected and that they are responding 'blindly' to with no understanding of what they are doing and certainly no conscious thought that the possession of a wattle of just the right shade of red is a sign of health and a good augury for the future. Perhaps even the monkey grunts have a simple explanation. We are impressed because we cannot hear the difference, but this may just be because our ears are not initially attuned to the sounds and with a little practice we might be just as good. Being mystified by what an animal does is no guarantee that the animal itself is clever.

The charge of stupidity can only be met by showing that animals do more than respond in an automatic 'innate' way and that they are capable of learning to adapt their behaviour to the particular circumstances they are in. If they go beyond the automatic responses they were born with and show enough understanding of the world to be able to change what they do or manipulate it to their own ends, then that would be real evidence of the kind we are looking for. An animal that can learn to make life better for itself by changing its behaviour must have at least a minimal understanding of the way its world works.

There is undeniable evidence of ability to learn almost wherever we look. A flock of hens, scratching peacefully together, may not look like a hotbed of learning but, in fact, each hen has learnt, often painfully, exactly where it fits into the group. When hens are first put together, they fight. Neck feathers raised and heads in the air like fighting cocks, they deliver vicious blows to each other. Combs and faces may get covered in blood. But a few days later, everything has calmed down and after a few weeks, all is apparent peace and harmony, apart from the odd 'reminder' blow that a hen that regards herself as of high social status may occasionally deliver to those she sees of inferior social rank. The legendary 'peck order' of hens is a fact of life, at least in groups that are small enough for each hen to know the others in her group as individuals. Walt Whitman could not have been more wrong when he wrote:

> *I think I could turn and live with animals, they are so*
> *placid and self-contain'd . . .*
> *Not one is dissatisfied, not one is demented with the*
> *mania of owning things,*
> *Not one kneels to another, not to his kind that lived*
> *thousands of years ago . . .*

The placidity that so impressed Whitman, not just in hens but in many other animals too, is the result of a previous period of strife in which positions in a social hierarchy are often settled with brute force – the very opposite of Whitman's vision. In fact, hens, like other animals, spend a great deal of their lives giving way to others and forcing others to give way to them, The outward appearance of harmony has only come about because each animal has already learnt its place the hard way and learnt that there is more to be gained in the long run by passively giving way to a social superior than by continuing to fight the same bloody battles day after day with an animal that is likely to defeat it. Peace is bought at the price of each hen learning to be pragmatic and accepting her place in the hierarchy, even if it is a relatively low one. Hens get to know each other as individuals, recognizing the characteristics of their flock-mates' faces particularly by oddities of their combs and wattles and giving way to them over food or roosting sites if social position demands it. Learning here is a social necessity, an integral part of nearly every moment of every day.

And as for animals not having a dementia for owning things, again, it is easy to be misled by the outward appearance of tolerance and harmony and not to realize that a highly developed ability to learn may mask the underlying conflicts for ownership of food, safe nesting sites and, above all, land.

In fact, I hope the ghost of Walt Whitman will forgive me if I say that 'being demented with the mania of owning things' is really rather a good description of the behaviour of many animals when they are defending their mates or a piece of land they regard as theirs. Hell hath no fury like a cockerel that sees another male attempting to mate with 'his' hens and, even when not engaged in physical fighting, animals spend a lot of time labelling what is theirs and proclaiming their ownership of a prized possession. All it takes in the winter is for a slight raising of the temperature and male great tits start giving their territorial defence songs, even though singing takes them away from the serious business of finding enough food to survive. The dawn chorus that delights us on spring mornings is really nothing more than birds warning their rivals to go away. 'Mine!', 'Keep off!' are their messages. This mania with owning things and telling the world about it is extremely time-consuming and exhausting. It is not as damaging as actual fighting because it is certainly better if a rival can be persuaded to leave through singing at him than having to resort to physical violence. But if song challenges are too frequent and if ownership proclamations take up too much of the day, then even these substitutes for real fighting begin to take their toll, if only in time that might be spent feeding or doing other things. And, just like the hens that learn to accept their place in the peck order without, in the end, the ruffle of a feather, so the owners of territories learn which song challenges are worth responding to and which can be safely ignored. By learning which individual bird is giving the challenge, they save themselves an immense amount of time and effort. Bruce Falls, working at the University of Toronto in Canada, has made a detailed study of the territorial songs of many different birds and has found that one species, the white-throated sparrow, has an extremely effective way of cutting down on the amount of 'rival-chasing' work it has to do. Males learn the songs of individual male birds and then selectively learn not to respond to the ones that pose no threat.

Small, rather nondescript birds, white-throated sparrows are indistinguishable from many other sparrow-like birds except for their clearly defined white throats and black-and-white striped heads. They are found throughout North America and the males have a distinctive territorial song that is sometimes described as 'Old Sam Peabody, Peabody, Peabody' uttered in a high clear whistle. All male white-throated sparrows sing this same basic song but each male also adds his own variation, a slight change in pitch, for example, while retaining the species-distinctive pattern. This means that a male's song can be recognized (by other males and other females) as being that of a white-throated sparrow, and can at the same time be identified as coming from a particular male.

In the spring the males set up breeding territories (each of about a third of a hectare), in which the females nest and rear their young. The territory serves as a haven and a food store and is thus extremely important to the male's chances of successful breeding, so that he will defend it vigorously. In suitable habitat, each territory will be surrounded on all sides by the territories belonging to other white-throated males. Now clearly, once the males have set up their territories and marked out all the suitable country, the only serious threats to a territory owner will be from landless outsiders that might try to oust them from their chosen patch. The neighbouring territory owners, in possession of their own resources for breeding, will be getting on with the business of attracting mates and helping them to rear young. There is no point, therefore, in a territorial male responding to the song of one of his settled neighbours. What he should spend his efforts on are the young hopefuls – males without territories of their own that are constantly on the look-out for one to take over. Falls found that white-throated males he was studying in Algonquin Park, Ontario, did exactly that. They quickly learnt the individual qualities of their resident neighbours' songs and did not respond to them even when the neighbours were actively singing in defence of their territories. But the minute a stranger arrived singing an unfamiliar song, the challenge was taken up, the owner of a territory would sing and, if necessary, chase the interloper away. Falls followed up his observations by making tape recordings of the songs of different males – those that lived close to a given male and also those of complete outsiders. The birds responded to the tape recordings as they would to real males (hardly surprising in view of the fact that sound is often the first indication of the presence of a rival hidden by leaves). They ignored recordings of their neighbours' voices and sang back at recordings of strangers. They evidently used their knowledge of 'safe' neighbours to cut down on the amount of singing they had to do – apparently regarding them as 'dear enemies'.

But then Falls discovered an even more interesting fact. If he played the song of a neighbouring male in the wrong place – on the south side of a territory, for instance, when that male normally occupied the territory to the north of the target bird – the response was as great as that to a complete stranger. The song was responded to with full territorial challenge. What the birds had learnt was evidently not just the individual characteristics of another male's song but *where* he was supposed to be. If he moved from his accustomed place (or appeared to move when the tape recorder was played from the wrong place), he was no longer regarded as the safe rival, content with his own patch of land, but a potential interloper on the prowl for a take-over. The dear enemy in the wrong place had become a real enemy, to be responded to as such and seen off. The peace established between territorial males was thus in this case resting on a complex learning process that involved the birds adjusting their behaviour not only to which individual bird they heard singing but whether that bird was in the place that it was supposed to be. We move even further from the idea that animals respond 'stupidly' and automatically to their environments.

Nevertheless, we could still say, both of the white-throated sparrows and of the hens learning their places in a flock, that the feats of learning and memory involved are impressive but not staggering. Given a little practice, we could probably recognize all 10 members of a flock of hens and even, straining our ears very hard, pick up the differences in the song of one sparrow and another. The differences are there: hens' wattles and combs do look different and a sound spectrograph will assure us that different male sparrows do have recognizable differences in their songs. There are some cases of animal learning abilities, however, that are of a different order of complexity altogether and, if we attempted to rival or better them, it would be we who would be made to look stupid.

Many animals, from foxes and squirrels to small birds like marsh tits and chickadees, hide food when it is plentiful and then come back and eat it some time later. In the case of a fox killing more gulls or chickens than it can eat at any one time and then burying the rest, the animal will remember perhaps two or three 'larder' sites. But marsh tits and chickadees may hide hundreds of food items in the course of just one day, each one in a separate place, and then find them again a few days later. During a whole year they may hide and then successfully recover thousands of items, each one carefully concealed behind a piece of bark or down a hole. (Some birds have been seen to drop seeds down hollow poles or to hammer them into wood so firmly that they could not get them out again.) This would suggest an astonishing memory for where food has been hidden (coupled with an occasional lack of realism about the practicalities of recovery). It implies that the birds can keep in their heads a memory for hundreds of different places where they have hidden food.

But before jumping to the conclusion that their memories are really that good, we have to rule out simpler explanations, such as that they locate the hidden seeds by smell or some other cue that we humans are unaware of. Perhaps when a seed is hidden, there are tell-tale marks such as disturbed bark, which can give a clue as to where the food is. If this were the case, the bird would simply have to remember roughly in which area it had hidden food and once it was there, be guided by on-the-spot signs it had left behind before.

However, a series of studies done by David Sherry and Sarah Shettleworth in Toronto and John Krebs in Oxford showed that this could not possibly be the case. They used captive marsh tits (which are European) and chickadees (North American birds which look and behave very like marsh tits, a fact which made the Toronto–Oxford collaboration so easy) and gave them artificial trees – tree trunks with holes drilled in them, highly suitable for hiding seeds in. Fortunately, at the right time of year, both marsh tits and chickadees will oblige by hiding large numbers of seeds in such artificial trees and they will still do so even if each hole is fitted with a little Velcro door that can be raised or lowered. It may sound rather strange to persuade birds to hide seeds in holes and then to put doors on the holes but there was a very good reason for it. Because the birds had to pull aside the Velcro covering in order to see whether anything was inside when a door was closed, they automatically gave a very clear indication of which holes they thought seeds were hidden in, even when they could not see them. Tearing aside a Velcro strip comes very easily to a bird that is used to tearing bark to get at food, so the trees with their little doors provided a reasonably naturalistic environment for them as well as allowing the experimenters to separate memory for particular holes from just going to a hole because they could see food in it.

The Velcro doors could be raised or lowered depending on

what the researchers wanted the birds to do. In one of David Sherry's experiments, for instance, there was a forest of artificial trees, each tree having several holes 0.5 cm in diameter and each hole could either be exposed (door up) or closed (door down). Sherry gave his chickadees some sunflower seeds to store and initially kept the little doors up so that the birds could put their seeds into the holes. He carefully noted which holes the birds used and then chased them out of the aviary where the artificial trees were. 24 hours later, he allowed them back in again but, in the meantime, he had closed all the doors so that now all the holes, both the ones with seeds in and the empty ones, were covered. Although the birds did not at first realize this, he had also removed all the seeds that the birds had hidden before he closed the doors so that not one single hole contained food. This meant that all the holes looked and smelt identical. The only difference between the holes was that the birds had used some of them to store seeds in the day before. Could the birds remember which of the 72 holes available to them in the trees they had actually used for the 15 seeds they had been given? The results were impressive. The birds systematically searched those holes they had put seeds in, even though they were now empty, tending to ignore the ones they had not used. By scoring which holes had their doors pulled open and comparing this to his record of which holes had been used for storage, Sherry was able to see how good the birds' memories were. They turned out to be excellent. The birds tore open the doors and selectively searched where they had put seeds the day before, even when there was no smell of food, sight of it or any other local cue to guide them as to what might be inside.

Sherry also showed that the birds could go one stage further and remember which holes they had visited during a search and that they took care not to revisit a hole once they had taken the food from it. To show this, he slightly modified the procedure for his next experiment. After he had allowed the birds to store their food and then kept them out of the aviary for 24 hours, he allowed them back but this time he left the food in the holes and kept the Velcro doors open so that the birds could see even from a distance which ones had food in them. He allowed them to eat half the food, that is, to visit half the holes they had filled the previous day and eat what was inside. Then he once again removed the birds and waited another 24 hours. The next day – now 48 hours after the original food-hiding had taken place – he removed all the remaining food from all the holes and covered each one with its Velcro door before allowing the birds back to see what they did. Now, faced with 72 Velcro-covered holes, the chickadees tore open only the doors of 'their' holes that they had not visited and plundered the previous day, leaving alone those holes that they had already removed the food from. They knew that if they had taken food from a hole 24 hours before, there was no point in visiting it again because it would be empty. They concentrated their attentions on those holes where (had it not been for Sherry unfairly removing everything) there would have been food. It would seem, then, that the birds could genuinely remember where they had stored food and, even more impressively, which holes they had subsequently visited and taken stored food from.

But there is one more important explanation that has to be ruled out before we can conclude that wild birds habitually remember the detailed positions of hundreds of items. It is just possible that birds do something similar to what a lazy motorist does when he can't remember precisely where he left his car but can vaguely remember that he left it 'under a tree' because he always tries to park it in shade. Such a motorist might give the impression of having a much better memory than he really did if he followed a preset rule of always parking under a tree. If he went straight to his car under the tree in the right-hand corner of the parking area, we might be very impressed because he seemed to have remembered which one, out of the hundreds of spaces available, he had chosen that morning. But if there were only four trees in the car park and he always parked it under that one if there was space available, what we would be witnessing would not be a prodigious feat of memory but the application of a simple rule 'Look under tree A, then B, C, and so on'.

If birds adopted similar rules – always hoarding food in certain favourite places and then looking systematically in those same places when they wanted to recover it – they could achieve a superficial appearance of having superb memories without in fact remembering all that much. This explanation, too, can be ruled out. Neither chickadees nor marsh tits do search systematically in the same way every time and they do not go back to the same places in the same order every time they want to find food. They move around, utilizing new places and then clearly remembering where they have stored things. They have no 'beaten track', no routine that could cut down on the amount they have to remember. And a recent Toronto/Oxford experiment shows that they can remember where food is even when they themselves have not hidden it, firmly ruling out the 'favourite parking places' explanation of hoarding and finding food.

This time, the birds studied were not chickadees but the very similar European marsh tits. The experiment itself was also based on the one that David Sherry had done before, with artificial trees, holes for storing seeds and Velcro doors over the holes. Only now, behind each Velcro door was a tiny clear plastic window. The trick was that, this time, humans, not the birds, hid the food. Into some of the holes, made inaccessible by the plastic window, were put bits of peanut. Initially, all the Velcro doors were left up so that when the marsh tits were allowed to explore the trees, they kept coming across holes full of peanuts but they could not get at them because of the plastic. After this, no doubt frustrating experience, the birds were removed from the trees and the Velcro doors were pulled down over all the holes. When the birds were later allowed back, they systematically started pulling up the doors of those holes in which they had previously seen peanuts. They clearly knew and had remembered for 24 hours which holes contained peanuts and which did not, even though they themselves had not put any food there. All they had had was a tantalizing glimpse of some food that they had not been allowed to eat, and yet this was enough for them to remember which holes might be worth investigating in the future.

The conclusion does, then, seem inescapable that these small birds have a phenomenal memory for places. In the wild, hundreds or thousands of sites are learnt about and the birds remember not just whether a hole once contained food but also, as we know from David Sherry's experiments, whether that hole is still likely to have food in it. They apparently keep their memories intact and up to date for hundreds of different items all at once, even after just one brief experience of seeing where something is hidden. A specialized, unusual sort of memory, perhaps, and one that we ourselves might not find particularly useful, but also one that cannot be dismissed lightly. The memories of these birds are prodigious. They leave us in no doubt that animals can learn and learn extremely effectively. By now, this much should be obvious.

But what we are going to do now is to look at an even more extraordinary phenomenon – an animal that not only learns for itself and profits from its own mistakes but one that is capable of learning from the experiences of other animals as well. It is a species where what has been learnt by one animal can diffuse through its whole social group by processes of copying and imitation. It even passes its knowledge from generation to the next, leading some people to say that this animal has the rudiments of 'culture' or 'tradition'. Whether or not we choose to go so far as to use these particular terms, it will be clear that words like 'innate' or 'automatic' are a wholly inappropriate description of much of the behaviour of this animal. Its apparent cleverness, cunning and ability to outwit the efforts of human beings to get rid of it have engendered exasperation and even, eventually, grudging respect.

There is a war against this particular animal, a war that has now been going on for hundreds of years and one that human beings have still not yet won. Many people would be glad to be totally rid of the animal in question but it is not so easy to eradicate such an intelligent opponent. As an inhabitant of sewers and other unsavoury places, it evokes revulsion. As a spreader of disease and spoiler of food, fear and loathing. Even its appearance – long, hairless pink tail and yellow teeth – is enough to repel many people. In short, rats do not at first sight seem likely torch-bearers of animal 'culture'. And yet, this is precisely what they seem to be. 'Culture' and 'tradition' are words that do force themselves, perhaps unwillingly, to our lips when we look at the organization of rat society and behaviour. Perhaps the fact that they are not most peoples' favourite animals will actually make it easier for us to be objective about their achievements. If 'surely not' rather than 'of course they can' is our first reaction, then we are possibly less likely to fall into the ever-present trap of anthropomorphism that awaits us with cuter or more appealing animals. So let us, as with the green blobs, make an effort to put aside our prejudices about appearance and life style and look beyond all that to the way rats run their lives and the explanation for their ability to make the best-laid plans of men go wildly astray.

The problem – or, depending on how you look at it, the opportunity for culture – comes from the fact that rats are highly social animals that often live where they are not wanted by human beings. So, the obvious thing to do is to poison them and this people have done with a vengeance, often using what are called anti-coagulant poisons such as warfarin that prevent the rats' blood from clotting and so making them bleed to death from internal haemorrhages (a particularly horrible death, incidentally).

One of the results of the mass-poisoning campaigns that have been conducted was that the rats became resistant to warfarin and other similar poisons – their bodies actually became able to cope with the poison so that they could eat it and not bleed internally. This happened because a small minority of rats had blood that would clot even in the presence of anti-coagulants and these rats survived better than the others and their offspring became commoner and commoner. Whole new generations of warfarin-resistant rats appeared, plaguing farmers who then demanded something better. So the scientists had to invent something else to kill the rats with, and then, when the rats became resistant to that, something else again. But while the new poisons were being developed and tried out – in some cases very successfully (at least at first) – something else was happening. It was not just the bodies of rats that had changed and were protecting them against the effects of poison, their behaviour changed too so that they became less likely to eat the poison in the first place. In some areas, such as the south of England, rats became almost impossible to control even with new and sophisticated forms of poison. On one Hampshire farm, a massive extermination programme resulted in only a third of the rats being killed in 14 days. Two-thirds of them were completely unscathed despite being surrounded by bait known to kill any rat that ate it.

Clair Brunton and David Macdonald of the University of Oxford then decided to put tiny radiocollars on some of the rats on this farm so that they could follow their movements and discover what they were doing. They found that many of the rats would run right past the poisoned food and not even taste it. They would then travel considerable distances and find food that was perfectly safe. Part of the explanation for this behaviour was that the rats were just extremely wary of anything new and avoided it. The poisoned bait had not been around before and so the more cautious rats would have nothing to do with it. They chose to travel further afield to feed on something that was familiar (and, of course, much safer). In this case, a poison-resistant rat was an ultra-cautious rat and there is nothing to suggest that they were being particularly clever about it. If some rats just happened to be more cautious in what they ate than others (and there are character differences among rats just as there are among other animals, so this is quite plausible), then in a world where new food was likely to be dangerous, the more cautious rats would be more likely to survive, reproduce and pass their cautiousness on to their offspring. Conservatism in this case pays and an appetite for novelty or too much curiosity about something new will almost certainly kill the rat. Such a process is the very opposite of learning. It looks as though what is going on is genetic selection on an innate tendency to be wary of novelty. Two versions of blind instinct – the cautious and the not-so-cautious – get tested in the dangerous environment of rat poisons, and the cautious version wins.

This is by no means the end of the story, however. The rats have many more tricks in reserve or, to be more accurate, a poison-resistant rat is not just a wary rat or a rat with a physiological resistance to poison. It is a rat that can register what is going on, and learn to adjust its behaviour accordingly by using every scrap of information there is to tell it what it is safe to eat and what must be avoided. To show how it does this, we leave the farmyard behind for a moment and turn to some experiments done in the laboratory on the white rat, the domesticated cousin of the wild Norway rats that are regarded as such pests by farmers. Bennet Galef of McMaster University has found that rats are capable of learning – and learning from each other. It is perhaps not surprising that rats, like many other animals, learn to eat more of food that tastes good and to avoid food that tastes bad or makes them feel sick afterwards. What Galef showed is that rats use each other as 'testers' (I wouldn't dare to use the word guinea-pig) for what might be safe or poisonous food for themselves.

Galef's method was quite simple. He would keep pairs of rats together in cages so that the members of the pair became quite familiar with one another. Rats are, even in the laboratory, very social animals and the two would be in the habit of regularly sniffing and grooming each other as well as sleeping in close physical contact. He would then take one rat of the pair out of the cage and, out of contact with its partner, give it experience of eating a particular sort of food and put it back with its companion to see what sort of effect it had on the rat that had been left behind. For example, he might take the rat out (he called it the 'demonstrator' rat for reasons that will become apparent) and give it food flavoured with a strong distinctive smell like cocoa or cinnamon, which neither of the pair had eaten or smelt before (some rats can be persuaded to eat new things sometimes!) The rat left behind (which he called the 'observer' rat) could not see or smell what was going on at the time, but when the demonstrator rat was returned to the home cage, the observer subjected the demonstrator to an intense bout of grooming and sniffing, concentrating in particular on its mouth and whiskers. After this sort of intimate interaction had been going on for 15 minutes, Galef then took the second rat (the observer, which had previously been left behind) out of the home cage and offered it a choice of two strongly smelling foods, neither of which it had come across but one of which (cinnamon, say) had been eaten by its demonstrator companion a short time previously. There was a very strong effect. Whatever food the observer rat had been eating, the demonstrator ate too – cocoa if the demonstrator had been eating cocoa and cinnamon if the demonstrator had been eating cinnamon. Since the observer did not see what the demonstrator had been eating, it must have smelt it on its mouth or breath. The advantage of 'copying at a distance' what another rat has been eating is, of course, that if that other rat has eaten something and survived, the food is unlikely to be poisonous.

Galef also showed the opposite effect – that observers would avoid food if their demonstrator smelt of it and appeared to be ill. Using the same set-up as before – that is, pairs of rats that knew each other well – he took one rat out and gave it some saccharine to drink, saccharine being a highly palatable food to rats, although non-nutritive. It was also, he made sure, a very familiar food to all the rats in his experiment and so the demonstrator rat drank it without hesitation. He then made the demonstrator rat feel temporarily ill by dosing it with a chemical compound called lithium chloride and put it back with the observer. The two rats interacted intensely as usual, by grooming and sniffing each other. Two hours later, both rats were, quite separately and out of sight of one another, offered more saccharine. The rat that had been dosed with lithium chloride not surprisingly refused to drink very much of the saccharine – the association between drinking this substance and feeling ill was too close in time for it to ignore the connection, even though in fact feeling ill had nothing to do with having drunk saccharine. This rat would, however, drink other solutions, showing that it was not just a generalized queasiness that was reducing its appetite but a specific aversion to saccharine that it had liked before.

What was really remarkable, though, was the behaviour of its partner, the observer rat, that had neither drunk saccharine that day nor been made ill with lithium chloride. It, too, was reluctant to drink saccharine even though on all previous occasions it had shown itself extremely willing to do so. Totally out of sight of the demonstrator, its behaviour mimicked that of its sick companion, in that it rejected a previously palatable food. It drank a little more of the saccharine than the observer rat did but not very much. Something about the smell and strange behaviour of its partner had put it off. In the real world, this behaviour would be highly adaptive, since an obviously poisoned rat that smelt of something distinctive could become an object lesson for all the others that whatever it smelt of was to be avoided. In the intimacy of their burrows, rats would have plenty of time to notice smells on each other. Galef's experiments show that they seem to be able to recognize the difference between a healthy rat (where what it smells of becomes a stimulus to be sought out and eaten) and a sick rat (where its smell becomes something to be avoided). This 'knowledge' of each others' state of health evidently takes some time for the rats to pick up. An observer rat has to be in the company of a sick demonstrator for at least half an hour before it shows evidence of avoiding what it had been eating or drinking. And the two rats have to know each other quite well for the observer to take any notice of the symptoms. Being put with a strange rat, even a severely ill one that smells distinctive, is not enough to put an observer off its food. But these conditions, too, are regularly met in wild rat colonies. Rats do know each other as individuals and do interact for long periods of time, giving themselves plenty of opportunities to assess whether there is something amiss with one of their companions and to use this information to keep themselves out of danger.

Clever, perhaps, but where in all this is the 'culture'? Information passing from one animal to another about what is safe or dangerous to eat is clearly an example of 'social learning' but 'culture' implies something more, the wisdom of one generation being somehow passed by non-genetic means to the next. The wariness a mother rat shows towards new kinds of food may also appear in her daughters but this can hardly be called 'culture' unless the daughters have learnt to be cautious by watching their mother. If she just passes on to them a genetic tendency to be wary, we are not looking at cultural inheritance at all. But if she learns something in her own lifetime and they in turn learn from her and put it into practice in their own lives, then at least by some definitions of the word, we are looking at a rudimentary form of cultural transmission. And on this definition, rats are clearly 'cultural'.

The particular problem that rats face – that of food that is possibly poisonous – makes them prime candidates for being able to learn from each other if they possibly can. The rat poisons that humans have devised are now so lethal that even small doses kill. No rat that eats them is going to get a second chance or be able to learn by taking a small sample what effect a new food might have. It will be dead. The trial-and-error strategy that another animal, a crow, say, faced with cast-off picnic litter with some unfamiliar food inside, can afford to take, is too risky for a rat for whom new may well mean lethal. On the other hand, old food supplies dry up and new ones must be found, so a rat that never tried out anything new might well starve. The risk of error in the trial-and-error process can be reduced by making use of the behaviour of other rats which, if they are still alive, must by definition have developed reasonably sensible eating habits. And if another rat has made an error and been poisoned, then the sensible thing is for the other rats to make use of its unfortunate experience and not make the same mistakes themselves. Young rats appear to adopt both these strategies. In wild colonies, young rats grow up with a strong preference for the foods their parents have been eating. They not only follow their parents to food sources and eat what they are eating, they are also strongly attracted to food that is surrounded by the urine, faeces and scents of other rats, particularly those of their parents. This means that even if the parents are not physically present, the young can be guided by the general rule that if many other rats have been there and have been eating for a long time, then the food is probably reasonably safe. This is not an infallible guide but it is certainly better than testing out every food as if no rat had ever tasted it before. It also means that information about what is safe for a rat to eat is transferred from one generation to the next as the habits of the parents are copied by the young.

We also know that information about what is dangerous can be transmitted in the same way. In one large scale experiment a whole colony of rats were regularly fed on two kinds of food – let's call them X and Y. Both X and Y were eaten by the rats and both seemed to be about equally acceptable to them. Then one day, food X was treated with lithium chloride, the substance that, as we have seen, makes rats feel ill for a while but does not kill them. It also has no taste of its own so that the rats unwittingly ingested it and those that did were made ill. The whole colony promptly gave up eating food X altogether and from then on only took food Y. Both foods were subsequently always presented without being contaminated with lithium chloride but even when food X had been lithium-chloride free for a long time, the colony did not go back to eating it. The effect persisted down the generations so that rats that had never had any experience with lithium chloride and certainly never known food X to be contaminated with it, refused to eat food X. A food that had once, before they had been born, been associated with a substance that had made some of their relatives ill was still avoided long after the rats that had eaten the unfortunate meals were no longer around. If such an effect occurred in a human society we would undoubtedly call it a cultural tradition and marvel at the persistence of a custom or taboo passed down by a social rather than a genetic route. In principle, although most human traditions tend to be more complex (although sometimes less rational), rats could be said to have their cultural taboos too, and for rather better reasons.

The fact that rats have not yet been defeated in the poison war being waged against them by modern chemical weapons is in large measure due to their ability to learn from each other and to pass on information about what is safe and what is dangerous to their offspring, which might otherwise have to discover everything for themselves the hard way – which in this case might prove lethal. As the war has been stepped up, their bodies have become resistant and their behaviour has changed so that they are protected by caution and conservatism. But resistance of the magnitude we see is only possible because, in addition to these basic defences, rats are highly intelligent animals, capable of picking up intimations of danger and changing their own behaviour accordingly. The fact that they can take notice of what happens to other rats and transmit what they know to young rats in the next generation makes them formidable opponents indeed. If we do not concede that they have a 'culture', then that must be because we have moved the goal posts and redefined what we mean by culture.

By this time, I hope I have convinced even the most sceptical reader of two things. First, it should now be clear from looking at the behaviour of a variety of animals that we cannot dismiss them as having a simple, crude, view of the world, totally different from the complexity with which human beings see it. Second, we are not dealing with organisms that do nothing except follow 'blind' preset instincts with no capability for leaning or adjustment to circumstances. On the contrary, the examples we have looked at illustrate how 'like us' even the most unlikely animals can be in the knowledge they have about the world, the fineness of the discriminations they can make and the way they can learn for themselves and from each other. We may still be secure in the superiority of our own intellects, perhaps seeing other animals as 'like us' but only in some very remote way. Nevertheless, it is clear that we are not alone in being able to recognize other members of our species as individuals, assess our chances for the future or anticipate what might be good or bad for our own well-being. Miss Halsey might by now be persuaded to move her foot just a little even if she had thought she wouldn't have to at all. But we do not stop here. There are yet more ways in which non-human animals are 'like us'. They show elements of what, if it appeared in ourselves, we would call 'choice' or 'decision-making', for instance. And they even appear to discriminate among different members of their own species on the basis of whether they judge them to be reliable, co-operative companions or cheaters, likely to let them down. In case this sounds too fanciful, however, too full of the anthropomorphism I was so keen to say was dangerous, let us look at two actual cases; one where an animal appears to 'decide' on the best course of action when several different ones are open to it and another where animals appear to weigh up the merits of their companions in a particularly down-to-earth way and give short shrift to those that attempt to grab too much for themselves. Then we can see whether words like 'decision-making' and 'co-operation' are justified or not.

Our first example is drawn from a garden, an ordinary town garden with a lawn and a cat. The owner of the house attached to the garden has thrown some pieces of bread onto the lawn, watched by the cat that has positioned itself under a bush some distance away. A single cock house sparrow alights on the fence surrounding the garden. He sees the food on the lawn but, having previously been chased by the cat, which he does not see for the moment, he waits on the fence and does not immediately fly down to the lawn. While he waits, he looks around and gives the throaty, gargling call that these sparrows so often give and that can be best written down as 'chirrup'. He gives it repeatedly, in a series of short bursts, scanning the garden all the time. After a few minutes, several other sparrows join him on the fence and eventually the whole group flies down and starts feeding on the bread, each bird continually looking around it between mouthfuls. The cat is a known hazard, long familiar to all of them, and has been known to appear suddenly from apparently cat-free bushes.

The whole scene could be said to be made up of a series of 'decisions' on the part of the individual sparrows. The first one to arrive spotted the food but did not immediately go to it. His initial decision was whether to fly down straight away, thereby keeping a large food resource all to himself or to wait until other sparrows had joined him first, which would mean sharing the food but would give him the protection of many pairs of eyes to look out for danger. Literally hundreds of studies have now shown that animals genuinely do find 'safety in numbers' because the more animals there are in the group, the more likely it is that at least one of them will be looking up and in the right direction if a predator appears. On the other hand, more pairs of eyes means more mouths, so animals have a problem in balancing the advantages of better advance warning of danger against the disadvantages of increased competition for food. Our sparrow, having had previous encounters with this particular garden and the particular cat that inhabits it, opted for safety. He did not risk flying down on his own but instead gave the chirrup call over and over again until he was joined by enough other sparrows. Mark Elgar, now studying cannibalism in Australian spiders but then doing his doctoral thesis on the sparrows on the roof of the Zoology Department in Cambridge, England, showed that the chirrup call is, in fact, a recruitment call and has the effect of making other sparrows come to the caller. The higher the rate of calling – the more chirrups per minute – the greater the number of birds that are attracted. He also found that the rate of calling was also related to the perceived danger of the situation. The closer he (Mark Elgar, substitute predator) sat to a food source, the more chirrups a hungry sparrow would give and the more companions he would have to have before approaching the food.

The sparrow on the garden fence had evidently 'decided' to wait until joined by other birds before risking this particular garden because he called at a high rate. Once the others began to arrive, he and they were then faced with the decision of when exactly to fly down and feed. When a certain number had assembled on the fence, they all decided to fly down together, but, even now, each individual still had moment-to-moment decisions to make about when to peck the food and when to scan the horizon for cats and other dangers. Elgar showed that each of these decisions was also affected by the precise situation the sparrows found themselves in. The less the perceived danger from predators (no cats or humans visible) the more likely a sparrow was to feed on its own without even bothering to wait until any others had arrived. The greater the danger, the higher the rate of chirrup calling there would be, the greater the number of assembled birds and the more looking around each sparrow did when it was feeding. He also found that when the need for food was most intense – during cold weather especially – the sparrows chirruped less and recruited fewer other sparrows before feeding. Evidently, when they are in most need of food, sparrows dispense with the inconveniences of having too many flock mates all trying to grab as much food as they can and get on with the serious business of eating. In this case, the 'decision' of whether to feed or to wait and recruit other birds gets tipped in favour of feeding, whereas in warmer weather, when the living is easier, the birds are that much more wary.

Elgar even found that the birds appeared to do a rough calculation as to whether the food they had found was or was not shareable with other birds. After all, there would be no point in recruiting lots of other birds to a food source if there were only enough for one when they all got there. He put out slices of bread, either broken up into several different pieces or the same amount in one single slice. He found that with the broken pieces, a sparrow that spotted them would chirrup and recruit other sparrows in the normal way. But, with the single slice, the same bird would appear much more reckless and, without waiting for any other birds to appear, would fly down and eat on its own. Being with other birds is such a mixed blessing that if there is only one item over which they are all going to squabble, it may be preferable for the discoverer to keep quiet about it and do without them, despite the risk.

The sense in which these sparrows are 'making decisions' is this. Each one has a series of options open to it – to feed or to chirrup, to chirrup at a high rate or a low rate, to join a flock or not, to peck at the ground or to scan for danger. These options are not pursued in the same way on each occasion. The same sparrow may sometimes do one thing and sometimes do another and we know from Elgar's work that which it does is determined by many different factors – the presence of danger, the number of other birds around, the ambient temperature, being just some of them. It even seems to depend on whether the food source is divisible into sparrow-sized portions or not. The sparrow appears to weigh up all these different influences and make a 'decision'. Sometimes priority is given to avoiding danger, sometimes to getting enough to eat, though there is usually an uneasy compromise between the two. Something inside the sparrow is balancing the pull of the sight of food against the push of the sight of a bush that has in the past concealed a particularly swift-footed cat.

As we have seen before, complexity does not in itself imply consciousness and it does not follow that the sparrows are rationally and knowingly balancing the likely consequences of doing one thing rather than another, but it does once again stop us from dismissing non-human animals as self-evidently so very much simpler than we are. Complexity is built in. The whole scene in the garden lasted only a few minutes and yet the sparrows made a number of decisions that could have had a major influence on whether they lived or died. During the course of a single day they might have had to make many other decisions, such as whether to fight, which mate to choose, what food to bring to the young, and so on. Some sort of weighing up of different courses of action is evidently part of their everyday lives, whether or not it is consciously done. And it is not just sparrows that have to choose between different options. The same is true of almost any animal at almost any moment of its life. Red deer assessing each other's fighting ability before a real fight, for instance, or birds on the point of migrating are equally 'making decisions' in the sense of choosing between different options or choosing the moment at which they will act.

Our second detailed look at an example of animal decision-making is at an animal that is much less widely known. In its tightly-knit social structure, we see what looks like a system of morality based on the principle of reciprocation of favours given in the past, coupled with sanctions against those that do not live up to their obligations. We are about to look at the far from legendary lives of vampire bats and the decisions that they make about each other.

Vampire bats, for all their ferocious reputations, are in fact highly social animals, at least to each other. They do, it is true, get their food in a way that is to us pretty repugnant – namely by drinking the blood of larger mammals, particularly domestic ones such as cows, donkeys and pigs. They alight, say, by the hoof of a resting horse, make a small, neat incision about 3 millimetres long in the back of its leg and take a drink of blood. Their saliva contains a substance that stops the blood from clotting so that they do not so much suck blood as lap it up by darting their tongues in and out of the wound. After only 15 minutes of feeding in this way, a bat may have taken in up to 40 per cent of its own body weight in blood – a large amount for the bat but insignificant for the horse, which may often make no attempt at all to shake it off. Or the horse may simply stamp its foot as though the bat were no more troublesome than a fly. After taking its relatively small quantity of blood (not the life-threatening quantities of the fables), the vampire bat flies off, leaving only a very small hole in its victim's skin. But it is this unusual method of getting its food (most bats feed on insects or nectar, not blood) that also makes the vampires unusual in their social behaviour.

Many species of bats roost together in large groups during the daylight hours in some cases for warmth, sometimes just because a particular cave happens to be dry and protected, and they all have the same idea as to what constitutes a good roosting place. Vampire bats, however, have an additional reason for staying together. There will be some nights when an individual flies out and fails to find any large animal suitable for it to feed on. It is soon in serious danger of dying of starvation. What is remarkable about vampire bats and must surely go some way to redeeming the reputation that exaggerated stories about their feeding habits have given them, is that in such circumstances, the bats will feed each other. A bat that has been lucky and had a good meal that night will come back to the place where the other bats are roosting and give some of the blood it has eaten to a hungry one. (The thought of how blood is regurgitated from the mouth of one bat into that of another should not make us lose sight of the fact that the act itself – the donation of food – is the equivalent of one of us giving up a meal we were going to eat ourselves to someone who needed it more.) The bats seem to know which of their companions is in need of food but – and this is the really interesting part – they do not feed just any bat that happens to be hungry. They preferentially feed relatives, such as their mothers and daughters, and they also feed particular unrelated individuals that they have consistently associated with in the past. They do not feed strange bats, nor do they feed every member of their own colony.

Of course, to give food to an offspring is not at all an unusual thing among animals as many species feed their young, often at considerable cost to themselves. But to feed a completely unrelated individual as these bats do is almost unheard of. What possible benefit could the feeder bat (the one that is after all cancelling out many of the benefits of its night's work by giving away much of what it has eaten) gain from saving another bat from starvation? Are animals altruistic after all?

Gerald Wilkinson decided to find out what was going on between the vampires. He spent many hours peering up into the recesses of hollow trees where vampire bats were roosting. (The effects of standing *under* a large roost of blood-eating bats and looking can best be left to the imagination.) He also established a captive colony at the University of Maryland where he worked so that he could keep a closer watch on particular individuals and which bats they fed and were fed by. He found that bats fed starving relatives but that they also fed certain other individuals, their 'roost mates' that they regularly roosted beside each day. These were not mated pairs because the closest feeding bonds tended to be between two or more females.

Wilkinson noticed that sometimes one bat would do the feeding and at other times it would be fed by the others, and it was this that gave him the clue as to what was going on. His studies on wild bats had shown him that getting – or missing – a meal on a given night was largely a matter of chance. It was not the case that there were some bats that were especially good at finding animals to feed on and always came back replete while others consistently failed to find anything. On the contrary, on some nights some bats were lucky and others were not whereas the position could be completely reversed a few nights later. This meant that there were plenty of opportunities for the bats to feed each other in a reciprocal way. On one night a bat might donate its food to another and a few nights later it could be repaid in kind. So when one bat donated a blood meal to a companion that had not eaten for some time, it was not benefiting itself in any way at that moment. It was clearly losing out, although if it were well fed, what it stood to lose by giving up some of its food was not all that great in comparison to what the starving companion, in desperate need of something to eat to avoid dying of starvation, would be gaining by having it. What the donor stood to gain only became apparent some time in the future. If its spell of good luck eventually turned into a run of nights when it found nothing to eat at all, then it could rely on the companion it had fed to rescue it from starvation by giving it some of its food. The benefactor would later become the recipient.

Wilkinson found that bats preferentially fed starving bats *that had fed them in the past*. They remembered which bats had been generous when they themselves had been down on their luck and those were the ones to which they gave a life-saving meal. They did not feed bats that had not helped them in the past. By maintaining strong reciprocal bonds with specific co-operative individuals, the bats guarded themselves against the vagaries and uncertainties of their peculiar feeding habits. And because any bat in the colony, even the canniest and most experienced hunter, could at any time find itself unable to find sufficient food, they all benefited from repaying their debts when they could and feeding their companions when they needed it. Any bat that attempted to cheat the system and take when it was hungry and never give anything back would, in the long run, be penalized. It might benefit once, but, the next time it was starving, the others would refuse to feed it and it might well die. Feeding another bat when it has plenty to give is the vampire's insurance against harder times in the future. The old adage 'keep your friendships in good repair' has a stark and pragmatic reality for these animals that may perish without it.

This example of co-operation among vampire bats is only an extreme case of a very widespread phenomenon among animals – interacting with other individuals to achieve a greater benefit to all of them than any one could achieve on its own. We saw another example in the garden sparrows, all benefiting from feeding together in a flock where they all look out for predators. Each bird is safer than it would be on its own and at the same time can eat at a faster rate because it does not have to break off its feeding so often to look around for predators. We see it in schools of fish, in pairs of horses whisking flies off each other's noses and in an oxpecker bird eating the ticks off a rhinoceros. But the vampire bats, with their knowledge of each other's past history, take social co-operation onto a new level. By giving food selectively to those animals that have proved to be reliable reciprocators in the past, they vastly increase the benefits that they themselves can derive from being in a social group. A cheating sparrow (one that fed all the time and never took its turn at looking round for predators) might well get away with it, to the detriment of all the others in the flock which would not be gaining anything from its presence, merely losing food. But a cheating vampire bat (one that never took its turn feeding others) could not. The individual recognition and vetting that the bats have evolved among themselves means that cheats cannot prosper and the benefits that an individual gains through feeding its partner all accrue to that individual itself in the future, not to other bats, or cheats of the group as a whole. The more it gives when it is needed by others, the more likely it is to gain when its own need is greatest. With vampire bats we see what look like 'sanctions' against those that do not co-operate and rewards for those that do. The individual and what it does begins to stand out from the crowd. The present is not all that matters and long-term bonds between individuals pay off in a future that may be some time away.

It is, in other words, hard to imagine anything further from the simple, automatic 'blind' instincts that we started out with. Perhaps they were a caricature of animal behaviour all along, or a straw man set up only to be shot down. I don't think so. Our knowledge of animal behaviour has grown immensely over the past 30 years or so and the more we have discovered, the more we have realized that the old view that animals just make simple responses to simple stimuli is incomplete and at best describes a few responses of some animals. Young animals, like newly hatched gull chicks, do respond to very crude cardboard models of their parents. And male sticklebacks do get excited by the most unlikely objects. But these are specific cases and not, we now know, the general rule. No, there has been a genuine change in our perception of animal behaviour brought about by detailed studies of animals that have revealed far more complexity than was initially imagined. And because this change is a relatively recent one, it has not yet filtered through to the outside world, and is not yet fully reflected in the views of those who are tempted to dismiss all non-human animals as necessarily simple or stupid.

So the purpose of this chapter has been to set the record straight. We have seen case after case where animal behaviour has turned out to be sophisticated and often more complex and baffling than even the people who were studying it had anticipated. We started with monkeys that can read far more into each others' grunts than humans ever could before the sound spectrograph came along to help them, and we have ended up with bats that keep tabs on each other. We have met birds with phenomenal memories and rats that pass the wisdom painfully gathered by one generation on to the next. To none of these examples have the words 'simple' or 'stupid' seemed at all appropriate, any more than 'blind instinct' or 'automatic'. So the first obstacle in the way of finding some definite traces of consciousness in animals that are not human should by now have been removed. If one of the hallmarks of consciousness is complexity of behaviour then at least many non-human animals must be considered as still in the running. They are 'like us' in unexpected ways – in being able to assess the prowess of other individuals, for instance, or in refusing to co-operate with cheats. And, however much we may resist the idea, these ways force us to keep our minds open to the possibility that the complex things they do are done because there are conscious minds at work behind them. Miss Halsey should surely have moved her foot quite a considerable distance by now if she would, as a result, avoid stepping on an organism that could do the things we have seen that many animals can do. And moving a foot is not really a very high price to pay for giving someone the benefit of the doubt – that is, if it avoids destroying a conscious being or a being that just might be conscious.

So let us suppose that Miss Halsey, confronted with the sort of evidence about animal behaviour that we have looked at in this chapter, actually does move her foot. A non-human animal has not asked her to do so in so many words but she is sufficiently convinced by its behaviour that it is 'like her' in ways that matter that she does so anyway. She thinks she recognizes the external signs of a mind inside the body she was about to step on and consequently she makes a detour around it. Our next question is the most difficult of all: was she right to do so or was her assumption that complexity of behaviour meant there was a conscious mind at work well meant but, in fact, erroneous?

Chapter Three

Bees do it

. . .the sending of the code was a reasonable
rational thing to do. It was a point of contact quite
unconnected with language, whereas the 'goodbye'
was only a superficial linguistic gloss.

Fred Hoyle, 'The Black Cloud'

CONSCIOUSNESS MAY GIVE RISE TO COMPLEX BEHAVIOUR BUT COMPLEX
behaviour does not, of course, have to come from consciousness.
This, you may remember, was the starting point of the last chapter.
Yet somehow the argument seems to have slipped a little and become
a bit woolly around the edges, so that now we seem to have reached
the point where complexity of behaviour in animals is leading us to
think that they might be conscious. Wrong. This is not what is being
argued and the previous chapter was really only a 'softening up'
exercise, designed to weaken the resolve of anyone who was going
to argue that non-human animals could not possibly be conscious
because their behaviour was so utterly unlike the only species we
know for certain to be conscious that the idea could be dismissed out
of hand. In hoping that I have achieved that, I must reiterate that
that is all I am claiming at this stage. The possibility of conscious
experiences in non-human animals is still just that – a possibility. In

this chapter, however, instead of going on to yet more wonderful and amazing examples of what animals can achieve, we are about to do an abrupt U-turn and to criticize the very evidence that might seem to help us in our quest for animal 'minds'.

I warned you earlier this would happen – that we would use evidence and then attempt to shoot it down. This may seem a crazy way to pursue an argument but it is the only way in which we will ever get to the really convincing evidence for consciousness in animals that we are looking for. 'Make the criticisms yourself before your opponents make them for you' would be one way of putting it. I prefer to say that we will base the case for consciousness in animals only on evidence that can stand up to real scrutiny. If we are prepared to be ruthless in what we accept for evidence in the first place then the slimmer volume of evidence we will be left with will carry a great deal more weight in the long run.

We are going to see that the well-intentioned Miss Halsey would, without further evidence, be in serious error on two separate grounds. She would be wrong to assume that complex behaviour, such as talking, necessarily implied complex processing going on inside the brain of the insect and she would be further wrong to assume that complex processing inevitably meant conscious awareness. If these distinctions are at the moment obscure to you, please don't give up. All will become clear as we continue.

We will begin with Miss Halsey's first potential error and make our particular target in this chapter the kind of woolly thinking that leads people to accept without criticism the conclusion that because an animal's behaviour gives an initial impression of being complex it really is complex and the animal is as 'clever' or as insightful as it appears to be. This error is in fact one of the enemies of the case for conscious experiences in animals although it may appear to be its friend because it weakens the argument in the long run when its flaws are finally discovered. We will look at the various forms that this woolly thinking can take and see how we can best eliminate it from the available evidence.

Appearances, we all know, can be deceptive. Magicians

specialize in the art of deceiving us, creating an effect that something has happened 'by magic' when we know there must be some trick or explanation that is quite compatible with the ordinary laws of physics. We know perfectly well that women cannot be cut in half and put back together again, but it can look extraordinarily likely that this is what is happening in front of our eyes. So even when we are fully aware that we must be being deceived, we can still be impressed and half believe what we are seeing. How much more gullible we become when no such obvious deception is being practised on us and how much more likely we are to accept what we see and hear if part of us already wants to believe that an animal really is doing something extremely clever. If no-one is deliberately playing tricks on us, why not accept what we see at face value and assume, for example, that an animal could be consciously working out its next move, taking a moral stance by giving food to a friend or calculating the answers to mathematical equations? And does it really matter anyway? If it seems to be doing these things, why not give it the benefit of the doubt and assume that human-like end-points mean human-like thoughts and feelings underneath?

There is a very good reason why it does matter. If an effect is achieved by following some simple rule then there is no need to invoke complex explanations, let alone conscious experiences, for what the animal is doing. But if the complex effect is achieved in ways that can be said to be 'like us' when we do the same things then conscious experiences in that animal become one step more likely. And, in order to be sure that it is 'like us', we have first to be sure that other simpler explanations can be ruled out. To say that it did not matter either way would be like saying that it did not matter whether the effect of a woman being sawn in half was achieved by magic or by a sophisticated trick. If we decided it was achieved by genuinely cutting her in half and sticking her together again, then we would at the same time have to accept that a major suspension of the known laws of physics and biology had occurred. But before we do this, we should make sure that simpler explanations, such as there were two separate women and a sliding box in the first place, could not give just as plausible an account of what is going on without the need to throw over a major part of our beliefs about the way the world works. We should start with the simplest possible explanation and see how far it can get us before invoking more complicated ones. This, at least, is a guiding principle for most working scientists who refer to it as Occam's razor. 'Entities should not be multiplied without necessity', said William of Occam. What this means is that we should always start with the simplest possible explanation and only when this has been shown to be quite inadequate should we move on to a more complex one. The history of animal behaviour is studded with cautionary tales of those who made startling claims for what animals could do and then had their theories punctured by a slash from Occam's razor. The most famous of these cautionary tales centres around a horse called Clever Hans.

The original Clever Hans lived at the beginning of the twentieth century but his many successors are still with us today. Clever Hanses frequently appear in circuses or on television shows. They are usually horses or dogs that are said to be able to count or even do mental arithmetic. Someone – a chat-show host or a member of the public – will ask the animal a question, such as what the answer to 7 + 9 is. If Clever Hans is a dog, it will bark its answer, whereas if he is a horse, it will stamp its hoof. When 16 barks or hoof stamps have been given, it stops. The audience is amazed. The animal must have worked out the answer in its head. How clever.

Clever Hans the First belonged to a German showman called von Osten who earned considerable sums of money through the supposed mathematical genius of his horse. He gave public demonstrations of what his horse could do that attracted such publicity that the scientific establishment eventually began to take notice. The startling idea of a horse that could do calculations, read, spell and understand musical intervals so intrigued Professor Stumpf, Director of the Psychological Institute of the University of Berlin, that he suggested to one of his students, Oskar Pfungst, that he should look into the matter and find out what was going on.

Pfungst quickly became convinced that the horse was not doing mental arithmetic or even counting but was instead responding to some cue, possibly quite inadvertent, from his owner. The problem was to show this. Pfungst suggested to von Osten that the horse's abilities should be studied under carefully controlled conditions and, so convinced was von Osten that his horse could really do what was claimed for him, that he readily agreed. Pfungst's test was to ask the horse a question to which his owner either didn't know the answer or had been deliberately given the wrong one. Under these circumstances, Clever Hans invariably gave the wrong answer. He could get the answer right only if his owner also knew the right answer and was with him at the time. It seemed that, although he was apparently completely unaware of it, von Osten was giving out some sort of cue to the horse when he had got to the correct number. So if the horse was asked to count to 7, he would start striking the ground with his foot and then, when he stamped this number of times, a slight sign from von Osten told him he had done enough and should stop, giving the impression that he had done all the counting himself. Von Osten vigorously denied helping his horse in this way and, indeed, to a casual onlooker, he had done nothing at all. The horse must have been responding to something almost imperceptible, a slight involuntary relaxation or a tiny expiration of breath perhaps. But there it was. When Pfungst arranged that von Osten gave the cue at the wrong place, the horse gave the wrong answer and when von Osten knew the answer and gave his cue in the right place, Clever Hans got the answer right too.

Then Pfungst noticed that von Osten made a barely discriminable downward movement of his head every time the horse began tapping and an equally small upward jerk when the correct answer was reached. It seemed to be these movements, quite inadvertent on the part of von Osten, that the horse was responding to, although the animal was so sensitive that tiny movements of von Osten's eyebrows or even dilation of his nostrils were also good enough on their own for the horse to get the right answer. The part of this story that I like best is that Pfungst, the sceptic, who knew perfectly well that Hans could not really do sums at all, could not himself help giving out cues that the horse could pick up. So if he (Pfungst) knew the right answer, Clever Hans got the answer right as well. Hans may not have been a mathematical genius but he was undoubtedly extremely 'clever' at picking up subtle cues from a variety of people.

This should not surprise us. Animals are constantly picking up cues from each other and become very adept at responding to cues both from their own and other species. Indeed, their lives may often depend on their being able to do so. Hans Kruuk, who spent three and a half years studying spotted hyenas in the Serengeti National Park in Tanzania, was very struck by the way in which wildebeest, gazelles and zebras would sometimes take no notice at all of hyenas passing very close to them. At other times, the same animals would become extremely nervous in the presence of those same predators and run in panic even when they were a long way away. It was clear that they were not simply responding to the mere presence of hyenas but to some minute detail of their behaviour that indicated that they were in danger – and even which species was most at risk. Kruuk found that hyenas had definite, short-term feeding preferences so that on one day they would hunt wildebeest and ignore zebra and on the next they would do exactly the opposite. The various prey species appeared to be sensitive to these switches and adjusted their own behaviour accordingly. When the hyenas were after wildebeest or gazelles, the zebra would be quite relaxed even before the hyenas' true intentions were apparent, whereas on another, 'zebra' day, they would be much more wary and jumpy. Some of the cues they used were probably connected with the numbers of hyenas there were in a group (hyena hunting groups are different sizes depending on what they are preying on) but this did not account for all the variation in their response to the hyenas and Kruuk was unable to say exactly what it was they were responding to.

The zebras, in other words, seemed to know beforehand how much at risk they were. They were not going to have their valuable eating time interrupted by predators that were not after them at all and posed no real danger. No point, for them, in fleeing headlong every time a predator came in sight, regardless of whether it had designs on them or not. Only if the threat was real was it responded to. But this kind of 'dicing with death' requires skill and, above all, it requires a detailed knowledge of predators and what they are likely to do next. The very subtlest cues – speed of movement, direction of gaze or posture – may be the give-away. But as long as these are reliable, life-and-death decisions (such as whether to stay put or disappear over the horizon) can be made using them.

Zebras can read hyenas like a book. Horses, applying the same skill, can learn to read human signs that humans themselves are unaware of. No wonder, then, that the shadow of Clever Hans hangs over all studies of animals where a human being is present – which means a high proportion of all observations and experiments that have ever been done in a laboratory and a good many done 'in the field' as well. If someone is training an animal, for instance, and rewards it for one action but not for another, then the animal may learn to take its cue from what the human has decided is the right answer. It was, after all, very difficult for Pfungst to show that Clever Hans was not really counting. Von Osten genuinely believed he was giving nothing away and even Pfungst himself could not help doing whatever it was that Hans needed to tell him what to do. It is not wilful, deliberate fraud we are talking about, but inadvertent betrayal of ourselves done with the best will in the world not to give anything away. And if a horse or a dog or a chimpanzee or even a rat can read the signs aright and so get a reward or avoid a punishment, there is an ever-present danger that it will do so and so make itself look much cleverer than it really is. This unfortunately seems to have been what happened in a long-running and bitter dispute over whether chimpanzees can or cannot learn a human language.

During the 1970s and early 80s some very startling claims were made for the 'linguistic' abilities of chimpanzees, some of whom had been taught human sign language (used by deaf people) or to manipulate symbols or signs in 'language-like' ways. The great excitement caused by these claims stemmed from the fact that language had come to be seen as the last great bastion of human uniqueness, the one thing that separated us from all other species.

Man the tool-user was no longer unique because many animals including the woodpecker finch (that uses a twig to prise an insect out of bark) and sea otters (that use stones to break open molluscs) were known to use tools. Man the tool-*maker* was not unique either since Jane Goodall's description of wild chimpanzees tearing the leaves off branches to make a rod for fishing termites out of their nests had shown that these animals were not just using but also preparing their tools as well. But man the user of language still seemed to be on his own. Here was something definite that separated us objectively from all other animals, an intellectual ability that outstripped all the rest not just in degree but in kind. So, when Allen and Beatrice Gardner of the University of Nevada claimed that a young chimpanzee called Washoe had not only learnt to use 132 signs of American Sign Language (used by deaf people and known as Ameslan) but had also been able to put them together in her own unique combination and to use them in meaningful and often novel ways, they struck at the very basis of what many people saw as the essence of humanness. And when the Gardners claimed that Washoe had a linguistic ability comparable to a human child in the early stages of learning to speak (9–24 months), their assault seemed to deliver a mortal blow to the comfortable idea that the rest of the animal kingdom could be kept permanently at a safe and respectful distance by virtue of its total lack of true language.

Certainly, Washoe's achievements, like those of other apes that have been taught versions of human language, are impressive. Washoe, raised in a highly social environment in which the humans around her not only constantly signed *at* her but also communicated with each other when in her presence in Ameslan, grew up able to use many of the gestures that deaf humans would use. She knew the signs for such things as 'eat', 'drink', 'tickle' and 'toothbrush' and would use the sign for 'open' in a wide variety of situations – for example, opening doors, opening boxes or turning on taps. Some signs she used in quite correct ways to refer to general categories of things. For example, she gave the sign for 'dog' when she saw a real dog, whatever the breed, to a picture of a dog or even just to the sound of a dog barking. The Gardners attached particular importance to Washoe's ability to put signs together in novel combinations to make a new compound word that she had not been specifically taught, suggesting that she really understood the meaning of the signs she was using. One of their most famous examples was Washoe's use of two signs, that for 'water' and that for 'bird', to make 'water bird' the first time she saw a duck swimming on a lake. Washoe also invented her own sign for 'bib' by drawing the outline of a bib on her chest.

Washoe, although a pioneer in her own way, is certainly not the only ape who it is claimed has crossed the human language barrier. David Premack, working at the University of Pennsylvania, used a different technique and taught another young chimp called Sarah to use plastic symbols instead of gestures. These symbols, of different colours and shapes, were backed with metal and could be stuck onto a magnetized board or moved around by either chimp pupil or human trainer. Sarah would learn that one symbol (deliberately chosen so that it did not resemble a banana in colour or shape) meant 'banana' while others, also chosen so that they bore no physical resemblance to the objects they were supposed to depict, meant 'apple', 'chocolate', 'Jim', 'Mary', 'Sarah', and so on. Other signs represented actions like 'give' or 'insert' – so that Sarah could construct whole sentences by putting the symbols for 'Jim' 'give' 'chocolate' 'Sarah' in the right order on her board. Some of the sentences that Sarah could obey had a complexity that seemed to put her understanding of them way beyond simple learning of order and into the realm of knowledge of the hierarchical structure of language – into the understanding of grammar and syntax, in other words. For example, when a human trainer put onto the board the sentence 'Sarah insert banana dish apple pail' Sarah would correctly put the right fruit in the right container, demonstrating, so Premack claimed, that she understood the grammatical construction of the sentence and knew that the verb 'insert' referred to both the banana into the pail and the apple into the dish. In order to carry out the command correctly, he argued, Sarah would have to go beyond mere word order (the word 'insert' referring to the word immediately following it and also to one several words later in the sentence) and to realize there was a carry over to some other part of the sentence.

So it seemed in the 1970s as though several apes were hammering on the door of humanness, having fulfilled at least some of the criteria for true language and thereby having broken down the last remaining barrier between 'them' and 'us'. They could use symbols to refer correctly to objects that either were or were not immediately present, or even as verbs. They could put their new-found knowledge to their own use and make novel combinations, thereby fulfilling the criteria of 'open-endedness' or 'creativity' that linguists kept insisting was one of the differences between human language and animal communication. And they apparently understood the rudiments of grammatical construction of sentences. Wasn't this enough? What more defined a human language? (It has to be said that this whole debate was made much more obscure by the lack of a clear definition of 'language'. The definitions kept shifting as people frantically tried to draw distinctions where up to now none had really been needed. Human language had been so vastly and obviously different from anything else that detailed criteria of what these differences might be were relatively new.)

But then Occam's razor began to take its toll. The Clever Hans effect (the chimps somehow taking their cues from their human trainers) or other much simpler and, let us be honest, less exciting, interpretations of chimp behaviour began to assert themselves. Perhaps the chimps really had no true linguistic abilities at all. Perhaps the woman had not been sawn in half. Perhaps what we were witnessing was nothing more spectacular than that which could be accomplished by a clever dog or horse picking up signs from its owner and acting on them.

There are three sorts of simpler explanations, three slashes from Occam's razor that we have to consider, and they are important because they potentially apply not just to chimpanzees learning (or not learning) human language but to all examples of animal behaviour where claims are made for cleverness or human-like abilities. These are the Clever Hans effect, which we have already discussed, the failure to specify what is expected before an experiment (so that the surprisingness of the actual result can be judged properly) and a lack of consideration of 'rules of thumb' (simple, unspectacular explanations for complex phenomena, such as sliding boxes for sawn-up women). We will discuss each of these in relation to the ape-language debate because this illustrates particularly well how misled we can be if we are not on our guard. The lessons to be learnt from the chimpanzee study have a very general applicability to other species that should carry over to all the other examples of 'clever' or 'complex' animal behaviour as well. This is not to say that the examples we have so far considered – the food-hoarding birds for instance or the food-sharing vampire bats – are *necessarily* flawed. (Remember how careful David Sherry had to be in showing that his birds really did remember the location of thousand of seeds and weren't doing something much simpler, such as finding them by smell or by always looking in the same places.) But it does point out how carefully all supposed evidence has to be looked at.

So if you are ready for some sodden blankets and cold water, let us see how well the claims made for chimpanzees and their linguistic abilities really do stand up to scrutiny. If this sounds disappointing and down beat (who wouldn't want to believe in Father Christmas or in real magic with at least half their minds) remember that it is all in aid of a better appreciation of the evidence for consciousness in animals that does remain. Evidence that can survive the onslaught of Occam's razor is indeed worth taking notice of. If it still remains when all attempts to fault it or substitute simple explanations have failed, it must be pretty good. So when we do look (as I promise we will in the next chapter) at evidence that seems to suggest higher mental abilities in animals, we can be reasonably confident that we are not constructing a house of cards. That is why I hope you will agree to being distinctly spoil-sport for the next few pages.

Now, the way in which the Clever Hans effect can potentially intrude itself into the chimpanzee studies will probably already be very obvious. The 'language' learning ability of both Washoe and

Sarah was dependent on their close association with human beings, both during their training and the subsequent language tests they were required to do. In Washoe's case, virtually all waking hours were spent in the presence of one or more human companions and her day was made as much like that of a human child as possible, with meals, bath, play and school – all with humans present. And when she was tested for her understanding and use of the various signs she had learnt, a human being was always there.

Sarah, although living in a laboratory during her adult life, was looked after in a human family for the first year after her arrival (at the age of about 9 months) and even later on her contacts with humans were extremely close and frequent. Her tests, too, were conducted with a human being there to place the symbols on the board and reward her if she did well. Ann Premack, who helped a great deal with Sarah's training, wrote (1976): 'People who raise chimps have high expectations for them as they have for their own children and when the chimps don't perform at these levels, the "parents" are often bitter . . . Aside from a human baby, I can think of no creature which arouse stronger feelings of tenderness than an infant chimpanzee. It has huge round eyes and a delicate head and is far more alert than a human infant of the same age. When you pick up a young chimp, it encircles your body with its long trembling arms and legs, and the effect is devastating – you want to take it home!'

So there we have it – on the chimp side massive opportunities for learning about human body language and on the human side a huge vested interest in their young charges performing well and doing the 'right thing'. The possibility of inadvertent cueing (and I do mean inadvertent – there is no suggestion of deliberate deception) must be taken very seriously indeed.

It was taken seriously by an American psychologist, Herb Terrace, who put himself through all the emotional trauma and difficulty of rearing a young chimpanzee and, teaching him sign language, and then had the objectivity to conclude that cueing was almost certainly happening on a massive scale. Terrace's chimp was called Nim and the key difference in what Terrace did to what had been done before was that he made extensive use of videotape in both training and testing sessions. This meant that, regardless of what people had thought was happening at the time when Nim's ability to use signs was being tested, there was a permanent and unedited record of both chimpanzee and human trainer that could later be looked at again and again. What Terrace found was that Nim's behaviour was very dependent on what his human trainer was doing or had just done, even though the person in question had no idea that this was happening. By looking carefully at the videotapes, Terrace could see that when Nim used a sign, such as that for 'hug', in many cases the human teacher had already given a complete or partial version of the same sign a few seconds before. In fact, Terrace concluded that during his last year of training, only 10 per cent of Nim's gestures could be genuinely said to be spontaneous or entirely his own and about 40 per cent were direct imitations of what his teacher had just done or was doing.

Even more revealing were Terrace's analyses of what looked at first sight to be Nim's 'sentences', such as 'Give drink', which he produced far more often than the non-sentence 'Drink give'. Nim, however, had learnt that 'give' was an excellent way of getting what he wanted. If his teacher had just signed 'Nim want drink?', all he had to do was use his tried and tested sign for 'give' and add on to it the sign for 'drink' which his teacher had just given. What might look like a sensible two-way conversation with a question ('Nim want drink?') followed by an entirely appropriate answer ('Give drink') thus probably had a much simpler explanation. Terrace also argued that Nim was not exceptional and that the few available films of Washoe also revealed a similar dependence on what a human has just done. For example, in one edited sequence, he points out that Beatrice Gardner is shown saying 'What time now?' and Washoe interrupts 'Time eat, time eat'. This, of course, looks very impressive. But a longer unedited version of the same exchange shows that what really happened was that Gardner had begun by signing 'Eat me, more me', after which Washoe gave her something to eat. Gardner then signed 'Thank you' and only then signed 'What time now?'. The fact that Washoe followed with 'Time eat' is therefore hardly surprising, as her human companion had just used both those signs immediately before and been given food for doing so.

Now, as Terrace has pointed out, the fact that the chimps are clever enough to take their cues from their human trainers and work out that to use a gesture that they have just seen a human use is a good way to get what they want does not mean that they are incapable of gesturing correctly without prompting. All it means is that we have to be incredibly careful to make sure that they can achieve what is claimed for them without doing this. In other words, the fact that the Clever Hans effect can work does not mean that it always does, but it must be eliminated before the more startling claims for chimpanzees can be accepted.

So, just as Oskar Pfungst had many years ago revealed the effects of 'cueing' by making an animal perform in the presence of an ignorant or wrongly informed human, attention was then turned to doing something comparable with chimpanzees. How would they behave if they had to cope with a person who had no idea of what they were 'talking' about?

David Premack, trying to control for the Clever Hans effect with Sarah, used as a tester someone who was completely unfamiliar with her language of plastic symbols. The result was disappointing because Sarah's performance deteriorated markedly. In particular, she seemed unable to put symbols in the right order to make a sentence as she had been able to do up to then. Instead, she put them down in an apparently random order and then started rearranging them. As two critics of these chimp experiments, Jean and Thomas Sebeok point out, this is precisely what would be expected if Sarah was searching for unintentional signs from her experimenter: she moved them around until she found an arrangement that was 'acceptable' to her human companion.

Premack, like the Gardners, then tried to devise even more stringent controls to eliminate cueing – tests in which humans were either not present or were in ignorance of what the chimpanzee was seeing and so should have been able to give an unbiased account of what the animal did. We will not here go into all the details of these experiments or of the often acrimonious debates about whether all possible human cues had or had not been eliminated but one rather sad fact emerges: the more carefully controlled the experiments are and the more precautions are taken to eliminate the possibility of some sort of chimp–human interaction, the less impressive the performance of the chimps becomes. Certainly, they can associate certain objects or actions with gestures and symbols, but then dogs or horses can do as much and we do not instantly rush to label it as 'language'. The more extravagant claims for knowledge of 'sentence' structure, however, do not look nearly as impressive when subjected to the ruthless glare of a video camera. We have seen how gesturing on the part of the human (who may be completely unaware of what they are doing and utterly shocked at the suggestion that they could be 'helping' the animal in this way) is a particular danger with studies like the chimpanzee ones, where social interactions and 'getting on together' are part of the success of training the animal in the first place and where emotional involvement by both parties is probably necessary for the projects to work at all. But the Clever Hans effect, although extremely powerful and the very devil to eliminate completely wherever humans and animals are in contact with each other, is not the only problem we have to contend with.

There is a second and even more insidious way in which astonishing results of clever animals can be produced, arising from the extraordinary capacity of the human mind to see connections and coincidences where none really exist, just simply because we 'want' to see them. This is not, of course, altogether a bad thing. Throughout our evolutionary history we (and other animals) have benefited from seeing connections between different things or events – for instance, that touching a certain plant leads to itching, or that dark clouds mean that it will rain. Using these connections, and acting in anticipation of events rather than assuming that the world is randomly put together, is clearly better than blundering around and being constantly surprised (and damaged) by every new circumstance. But it does mean that we, in our efforts to see connections and make sense of the world, are likely on some occasions to 'jump to conclusions' and believe that events are linked when in fact they are not. Superstitions grow up in this way. If two events are observed to follow each other on one occasion, people will frequently act as though A always follows B. They will avoid walking under a ladder because bad luck is supposed to follow and so never give themselves a chance to discover if there is any true connection at all between where they walk and what follows.

In fact, as a species, we are so superstitious, so prone to finding connections if we possibly can, that a whole science called statistics has had to be developed to curb our enthusiasm (shared by scientists and laymen alike) to assume that two events are connected when in fact they are not. Statisticians have what they call a *p* (probability) value that is applied to the results of experiments or observations and which is really an indication of how excited we are allowed to be by a particular result. We might think that it was absolutely extraordinary that we should toss a coin three times and each time it came down heads. But the probability of this happening even with a totally unbiased coin, is actually 1 in 8 – really not such a coincidence at all and nothing much to get excited about. Equally, we might think it was an amazing coincidence that if you and I were both tossing coins, we should both throw heads, but in fact the probability of this happening by chance is as high as 1 in 4.

The statisticians' method of calculating the *p* or excitement values (the lower the *p*, the higher the level of excitement you are permitted) involves specifying *what else could have happened*. In the case of the single coin being tossed three times, seven other things could have happened besides the head–head–head that actually did occur. These other seven are:

Head – Tail – Head
Head – Head – Tail
Head – Tail – Tail
Tail – Head – Head

Tail – Tail – Head
Tail – Tail – Tail
Tail – Head – Tail

Out of the eight possible things that could have happened, one actually did, so the probability of it happening by chance is, as we have seen, 1 in 8. With you and I simultaneously tossing coins, only four things could happen:

YOU	ME
Head	Head
Head	Tail
Tail	Head
Tail	Tail

So the probability of us both throwing heads has got to be 1 in 4. Fully half the time we should expect to have the same result because both of us throwing heads or both of us throwing tails constitute half the possible outcomes expected by chance.

What this means is that when we look at an animal behaviour experiment, and particularly one in which the experimenters want to believe that a particular result is extremely significant and says something dramatic about that animal's abilities, we must be quite sure that they have specified how likely that result was to have happened by chance alone because only then can we know how excited we should or shouldn't be by what has happened. Take, for example, David Premack's experiment on Sarah. He claimed that because Sarah could act appropriately when instructed to 'insert banana pail apple dish', she understood the hierarchical or grammatical structure of the sentence. Sarah was provided with a choice of fruit and a choice of containers but, as Herb Terrace has pointed out, she was only required to do one sort of action (put a bit of fruit into some sort of container for the whole of one test session). So obviously, the word 'insert' referred to both the banana/bucket problem and the apple/dish instruction – it couldn't refer to anything else. In other words, the number of possible outcomes of this experiment was actually quite limited and so the probability of getting the observed result by chance alone was correspondingly high. Only if the instruction 'insert' had been one of several alternatives, such as 'put under' or 'withdraw' or 'wash', would it have really been remarkable that she took the word 'insert' to refer to both instructions. If Sarah could understand a complex sentence such as 'Sarah withdraw apple banana dish insert peach can apple box' or 'Sarah wash apple red dish insert banana round dish', then we could believe that she understood complex sentence structure and could realize that words such as 'insert' and 'withdraw' referred to words that did not necessarily follow them immediately but might have to be 'carried over' to a later part of the sentence. Here, the likelihood of her getting the right answer by chance would be very much less and consequently, if she could do it, we ought really to get very excited indeed. If she could perform reliably on a variety of such complex sentences, then it would be very difficult to explain her behaviour in any way other than by saying she understood which objects were to be manipulated by which verb, even in a hierarchically arranged sentence. As it was, with just a limited range of alternative outcomes possible, the fact that she performed correctly is really not very surprising at all.

Chance can have other pernicious effects on our interpretation of animal behaviour. Suppose that a chimpanzee is using sign language or arranging plastic symbols on a board and most of the time what it is 'saying' doesn't make much sense, but just occasionally it produces a recognizable, human-like sentence. We would probably be tempted to be very impressed and say that the animal understood what it was saying. But this would, of course, be completely incorrect. What about all the other times when it babbled away and produced nonsense? If every time it strung words or symbols together a beautiful sentence resulted then, yes, certainly we should be impressed, and rightly so. But if most of the time there was nonsense and just occasionally something that made sense, then this is no more impressive than the occasional run of heads in a sequence of coin-tossing.

Of course, it is impressive if a patch of order emerges from chains of randomness and this will be the sequence we notice and remember. We conveniently forget the majority of occasions when there was no order or the animal made mistakes. The one word produced by the proverbial monkey at the typewriter will stick in our minds in the way that the rest of his random outpourings will not. 'A word!', 'Two words together!', we will say, attempting to read significance into the message and ignoring the screeds of paper covered with jumbled incoherent letters that we had to scan through before getting to our supposed message.

Our brains have been evolved to detect pattern and detect it we will even if it isn't there. The many people that have fallen to Monte Carlo fallacy and believed that there were predictable patterns in the random throws of dice or wheel are testimony to our determination to find order and predictability even where none exists. So here is another reason why videotape or film records of what chimpanzees do for whole sessions are essential. They guard against the temptation to pick out the good bits and ignore the times when the animals make no sense at all. They give us our statistical base from which we can judge the surprisingness of what they do. How often did Washoe sign 'water' and 'bird' separately, so how likely was it that she would have put these two signs together (as 'water bird' or 'bird water') anyway? How reliably did she sign 'water bird' when shown ducks or swans? These are questions to which we will never know the answers (easily portable video cameras were not available in the 1970s) but they serve to illustrate the kinds of questions that have to be answered before we should draw too many conclusions about the intellectual abilities of apes. That is not to say that they don't have them, but that in order to show them we have to be sure that they are doing more than Lady Luck could come up with by herself. Like the Clever Hans effect, chance results are what have to be eliminated before we can be sure of what we are dealing with. Very often they can be, in which case we can proceed with confidence. But if they are not even looked at, then we leave ourselves vulnerable to two easy slashes from Occam's razor. A particularly dismal kind of thud occurs if the supposed complexity of animal behaviour turns out to be due either to the animals taking their cues from humans or to a chance association that could have arisen by the ordinary laws of probability. But the let down is only slightly less embarrassing if, when all precautions to exclude these two possibilities have been taken, a third stroke of Occam's razor reveals that yet another kind of simple explanation fits the facts just as well, if not better.

The third category of potential fallacies arises from yet another of our human tendencies – that of assuming that a wide range of inanimate objects from talking dolls, ships and computers to mountains or the sea have powers of thinking and feeling. We all too readily attribute to them human-like properties of awareness as well as slip into thinking that they are somehow alive. In fact, we have a tendency to do the very opposite of what William of Occam said we should do: we assume a complex hypothesis (the ship is alive and sentient) rather than the simpler (that it isn't). Not that we do this consciously. We would probably deny it altogether if confronted with the charge that we really *did* think a car or a ship were conscious, but part of us, almost without thinking, often does make some such assumption. And the problem is that when we are confronted with an animal doing something complicated and apparently clever, this unconscious assumptions spills over into our interpretation of the animal's behaviour. The problem is compounded by the habit that biologists have recently got into of using terms that by their very nature imply complexity and even consciousness. Words such as 'assessment' and 'decision-making', which I used in the last chapter, occur commonly in the scientific literature. They are used as technical terms but they are clearly borrowed from everyday speech and still carry implications of conscious thought. If you look at the fine print you will notice that biologists usually cover themselves by saying that it should go without saying that when they use these words in a technical sense they do not mean to imply that the animals are necessarily consciously working things out or even doing anything particularly clever at all. The words are the same but the meaning is supposed to be different. Or, rather, the colour and dramatic impact of the words in everyday speech are retained (and how comparatively dull the animal behaviour literature would be if they were done away with) but everybody is supposed to understand that they have dropped their links with implied conscious experiences.

The trouble is that not everybody does. So when an animal is said to be 'assessing' another or 'making a decision' all the human connotations of these words come crowding in and the assumption that the animals are doing something very complex and human-like is made almost before anybody realizes what is happening. Simpler hypotheses are not even considered. If an animal is clearly 'sharing food' or 'penalising cheats', then it is all too easy to assume that what is going through its mind is the exact equivalent of what we would be thinking if we saw someone who was in need of food and altruistically decided to give them some of our own or that they shared our sense of moral outrage when a cheat attempts to get away with more than his or her due.

In their more sensible moments, biologists are fully aware of this and, while trying to keep the words themselves because of their dramatic effect but wanting to separate themselves from their implications, make it clear that many animals operate through 'rules of thumb'. Rules of thumb are simple laws governing animal behaviour stripped of any unnecessary connotations of complexity but which describe the actual way in which the animal is responding. So, for example, it might be possible to describe the female of a species as 'assessing' males (implying a complex, possibly conscious, evaluation of each one's worth) because she visited each male's territory in turn and then mated with only one of them. But it might turn out that the female's rule of thumb was simply 'mate with male with the longest tail' or 'loudest call', or whatever. All that would be implied here would be that the female had some way of responding to tail length or loudness of call, both of which could be done relatively simply. Provided that, say, tail length correlated well with other aspects of male quality (his health, vigour, lack of lice or other external parasites, ability to fight off other males, etc.), the biologist could maintain that the female was making a complex 'assessment' of the male's genetic quality but at the same time say that she was following the rule of thumb of 'mate with long-tailed male'.

Or, an animal might be described as 'making a decision' between eating and drinking, again implying a complicated (and conscious) weighing up of the likely consequences of various courses of action. But its rule of thumb might be 'stop eating when level of substance X in the blood reaches level Y and switch to drinking' – an automatic change no more complicated than the central heating system of a house switching itself off when a certain temperature is reached. We could, if we wished, describe a boiler as 'deciding' that a house was warm enough or an automatic car 'deciding' it was time to change gear but we would probably not do so because we know that the rules of thumb in these cases are so simple that a word like 'decided' is not really called for. And even if we did slip into using it, we would know perfectly well that there was not a conscious brain inside our gas-fired boiler. We would, say, just be drawing an analogy or using a figure of speech. In one limited respect the boiler behaves 'as if' it knows what temperature its house was at and had our comfort and convenience at the forefront of its 'mind'.

With animals, the use of analogy and 'as if' gets a bit blurred. Figures of speech subtly become hypotheses about what is really going on and what started out as a technical term borrowed from everyday speech slips back into its colloquial meaning. We are so ignorant in most cases about what the rules of thumb of any given animal might be using that we cannot with confidence say 'just a thermostat'. Instead, faced with the complexity of animal behaviour (and animals are genuinely far more complex than any man-made machine so far devised), we have a tendency to jump to the conclusion that it is much more complicated and mysterious than it really is. Because we don't understand fully how animal bodies function, we tend to assume that they achieve their complexity by thinking and working things out. But before we are entitled to conclude that that is really what they are doing (with all the implications that this might have for the minds that could be responsible for the complexity) we must be sure that what we are looking at could not be explained much more simply with a rule of thumb. A switch in a hormone level or a greater response to a long tail than a short one is a much simpler explanation of why a female mates with one male rather than another than would be implied by saying that she 'assesses every male in turn'. These rules of thumb can be very difficult to spot and very deceptive in leading us to think that something is complex when it really turns out not to be after all.

As a starting point and a way of keeping our feet on the ground, we can learn a lot from a very early computer program called ELIZA, which was designed to simulate the behaviour of a psychoanalyst listening to a patient. The dialogue might go something like this:

ELIZA	What would you like to talk about?
PATIENT	It's difficult to know where to begin.
ELIZA	Tell me the first thing that comes into your mind.
PATIENT	A garden in front of a house. My mother is picking roses.
ELIZA	Tell me about your mother.

and so on.

The important thing to realize is that such a program is extremely easy to write. Anyone with a working knowledge of a language like BASIC could make a passable (not indefinitely passable – the program would be 'rumbled' eventually) of an encounter between a psychoanalyst or a patient. All you need is a list of stock phrases, such as 'tell me the first thing that comes into your mind', to fill up awkward gaps in the conversation and an instruction to tell the computer to pick up certain key words such as 'mother' or 'father' and insert them into another stock phrase 'Tell me about your . . . '. Even with such very simple rules of thumb (single lines of program instructions) an eery feeling that there is someone in there listening to us pouring out our feelings can be created. Naturally, if the program is very simple it can be tripped up easily; for example, if we say 'My father died before I was born and I've never discovered anything about him' and the computer replies 'Tell me about your father', the illusion will be shattered. But even only slightly more complex programs could deal with this one and continue the illusion for much longer. We should never forget that even simple rules of thumb, as simple as a couple of lines of BASIC program, can very easily create the illusion of complexity and minds and consciousness.

But now to real animals. To be more specific to an insect, not to a hypothetical beetle that speaks English and startles unwary humans about to step on it but to a real insect familiar to almost everyone – the honeybee – that shows a quite extraordinary capacity to be 'clever' and to do the very things that if it were large and had a big brain we would suggest showed that it must be conscious. Not for the bee the close social contact with human beings that lets dogs and horses and possibly even chimps appear cleverer than they really are by picking up on human body language. Not for the bee the statistical sloppiness that an experimenter might slip into in dealing with an animal he was emotionally involved with and had already decided was conscious. No, if bees appear to be clever, that is because they do remarkable things despite the best attempts of humans to prove that they can't possibly be doing them. If bees were admitted to the circle of beings acknowledged to be conscious, then virtually the entire animal kingdom would be queuing up for admission too. Chimpanzees are one thing – they are so like us and so are dogs in their way. But an insect? With such a small nervous system and no real brain to speak of? It'll be jellyfish and cabbages next! It must be a trick or a mistake. But, as we shall see, it is not a mistake and bees turn out able to perform some of the most astonishing behavioural feats shown by any animal. By adjusting their behaviour to relatively simple stimuli and proceeding step by step (each one nothing remarkable in itself), the end result can almost be beyond belief. Bees have the rudiments of a 'language', they carry out 'assessments' of their world and they 'make decisions' of an extraordinarily sophisticated kind. They do it by following very simple rules of thumb and provide an object lesson to any one foolish enough to assume that such higher functions must necessarily be the result of a conscious mind. Bees and their behaviour are as much of an eye-opener as a lesson from a good conjurer would be if you could ever persuade him to give away the tricks of his trade.

The most widely known of bee achievements is, of course, their ability to communicate with other members of their hive about where food is. The Austrian zoologist Karl von Frisch shared the Nobel Prize in 1973 for showing that they did this through a special dance that a worker bee performs inside the hive. If a dish of sugar solution is put out near a bee hive, once one bee has found it then it is usually a matter of minutes before many other bees come to it as well. Something about the behaviour of the discoverer bee brings them there and that something is a dance which takes different forms depending on how far the new food source is from the hive.

If the food is close (less than 50 metres away), the returning bee performs a 'round dance', circling first left and then right on the vertical surface of the comb inside the hive. The other bees cluster around and then, stimulated by their contact with the dancer, fly out of the hive and search nearby. They apparently have little information about which direction to fly in and all the round dance conveys is the instruction 'search within 50 metres, although, helped by the smell of the food on the body of the dancer, they usually find the right place without difficulty. If the food is further away, on the other hand (say, over 100 metres), then the dance is no longer round but in the form of a figure-of-eight. The circling is interspersed with straight runs in which the dancer waggles her abdomen rapidly from side to side, giving rise to the name 'waggle dance'. This dance contains information both about how far away food is and also about which direction another bee would have to go to find it. The distance of the food is indicated by the general tempo of the dance (the nearer it is, the more rapidly the dancer dances) and its direction is given by the orientation of the bee during the straight run or waggle bit of the dance.

The problem the bees have is that they often dance on the inside of a dark hive where neither the food itself nor the sun (which is their major compass) is visible. Not only that, but they are dancing on a vertical comb when information has to be given to the other bees about which direction they should fly in the horizontal plane (food is unlikely to be found by flying straight up into the air). The bees cannot, therefore, indicate the direction of food by simply pointing or dancing towards it. They translate the flight path from hive to food (which will eventually be taken relative to the sun) into a direction relative to gravity inside the hive and the other bees re-translate this back into instructions relative to the sun when they get outside. So if the food is to be found by flying directly into the sun, the dancer will dance so that she does the straight 'waggle' run precisely vertically on the comb, whereas if the food is to be found by flying at an angle of 40 degrees to the west of the sun, she waggles 40 degrees to the left of straight vertical. She thus substitutes angle with respect to vertical for angle with respect to the sun and conveys, in the darkness of the hive, information to her companions as to the direction they should fly when they get out into the sunlight. The fact that they do this correctly shows that bees do indeed convey information to each other.

More recently, something even more extraordinary has been discovered about the bee dance. I mentioned that the vigour or intensity with which a bee dances is not a fixed quantity but varies in relation to the strength of the food it has found. So if, for example, food is very scarce, a relatively weak source of sugar or nectar will elicit vigorous dancing but if food is plentiful, so that source of food becomes not nearly as good *relative* to other richer food sources, then the same concentration of food will lead to less vigorous dancing. In other words, the bees act as if they know not only how concentrated their own food source is but how good it is relative to other places. How do they know what is going on at other food sites being visited by other workers? It does not appear that they visit many sites and then make direct comparisons. Von Frisch marked individual bees with paint and showed that most bees work steadily at their food source without taking a break to find out what is going on elsewhere. And yet somehow the bees accumulate at sites that have the richest food sources in the environment, implying that information of a comparative nature – which site is the best – somehow travels between them. An associate of von Frish, Martin Lindauer, showed that this was done through the medium of 'receiver bees' that stay in the hive while the foragers go out collecting. Because these bees receive loads from many different foragers returning from many different feeding sites, they are in a position to compare the sugar concentrations from all of them (a bit like a fruit merchant receiving apples from all over the country and comparing them for quality). Foragers returning with rich, sugary food are unloaded quickly but those returning with relatively dilute nectar arouse little response in the receivers and may have to search to find one willing to accept what they have brought back. The better the food, the more quickly the forager is unloaded. The forager is clearly sensitive to the response she is getting from her receiver because there is a direct effect on the forager's tendency to dance about her food afterwards. If the receiver unloaded her food quickly, say, in less than 40 seconds, then she is very likely to dance on the comb and to recruit other workers to go and visit 'her' patch of food. But if she found it difficult to find a hive bee to receive her load of food and the unloading process took 100 or more seconds, then she is unlikely to dance: 'her' patch had evidently yielded food that gave rise to only a half-hearted response on the part of the receiver bees, which, having experience of lots of bees from other areas, were effectively communicating that she could do better by going elsewhere. This system can obviously help each individual worker bee to collect food in the place that is most worthwhile. The richer the nectar coming into the nest, the choosier the receiver bees will be in unloading only workers with valuable nectar and the more likely a forager coming back will be to switch to another food site where the nectar concentration will be even higher than that which she has been able to find so far. But, by using the response of the receiver bees in this way, the foragers also make sure that, when food is scarce, poorer food sources – those that would be totally ignored when there was a lot of nectar around – are made use of. Foragers finding them will be rewarded by being swiftly unloaded, they will in turn dance vigorously and stimulate other foragers to go out and collect from the not very rich, but 'best of a poor choice' of food sources.

Now, we could describe all this by saying that the bees 'tell' each other where food is and 'know' which sites are the best, using terminology (which I did my best to avoid) that implies conscious knowing and conscious conveying of ideas. But the rules of thumb used by the bees are much simpler. They are things like 'Respond more to higher concentrations of sugar than lower', 'Dance with a probability proportional to unloading time', and so on. And yet, in conjunction, they result in a beautifully co-ordinated system, that gives the impression that the colony 'knows' where the best food sources are, and can track the best nectar around as new flowers open and others die. If we hadn't discovered the simple rules that govern this system, we might be tempted to read much more into it than is warranted. And if it hadn't been bees doing it, we might not have bothered to find out and assumed that things were more complex than they were.

In one sense (the technical sense, of course) the bees could be said to be 'assessing' their environment in that they appear to estimate which out of the many potential food sites open to them is in practice the most profitable. And there are other ways, too, in which the word 'assessment' seems just, if not more, appropriate when applied to bees as when applied to other animals, and when the bee hive, collectively, seem to take a 'decision' that looks suspiciously like democracy at work.

Honeybee colonies reproduce by swarming – that is, by the queen flying off with a large proportion of the work force to found a new colony, leaving behind the rest of the workers and enough young virgin queens to ensure that one of them will eventually emerge as the new queen. The days before her departure see some remarkable changes in the activity of the hive. The workers build special queen chambers, tiny inverted bowls made of beeswax, into which the queen lays eggs (20 or so) destined to become the young queens of the next generation. The workers feed the larvae that hatch from these eggs on special 'royal jelly' that ensures that they become queens not workers. (Both the queens and workers are female and which they become is determined largely by what they receive to eat in early life.) At the same time, the queen herself loses weight. She is fed less and less by the workers (previously she has been the object of their constant care and attention) and is constantly shaken and buffeted by them. The workers will push their queen with their front legs and shake her. This forces her to keep moving around the colony and, eventually, as a result of reduced feeding and this forced exercise she loses up to 25 per cent of her body weight. When the first young queen hatches, the old queen leaves the colony taking with her about half the workers (usually 10,000–15,000 of them if it is a strong and vigorous colony) and forms a swarm in a nearby tree. At this point, the queen and her swarm have no clear destination. They have left the old nest and have no new home and it is now that the most extraordinary decision-making process begins to take place.

Leaving the swarm of bees, which looks like a strange living beard hanging in the tree, some of the workers which had previously been foraging for food for the hive before the momentous event of swarming took place, fly out and start exploring the environment. Whereas previously they had been attracted to brightly coloured flowers and the smells of nectar and pollen, now they look for dark places – holes, cracks in tree trunks or caves in rocks. What they are looking for now is a place suitable for the colony to make its new home.

If a scout bee finds a suitable hole, she will spend a long time, perhaps an hour, systematically investigating it. She looks at the outside, hovering all around the hole, apparently inspecting it visually. Every so often she goes inside the hole and walks around the cavity, at first close to the next entrance and then further away from it until eventually she has walked around the entire inner surface. Thomas Seeley, of Yale University who has made a special study of this behaviour, estimates that a single bee may walk a total of 50 metres in this way and by making bees walk in artificial cylindrical nests that could be rotated (a sort of bee treadmill), he showed that the bees estimate the volume of a cavity by how far they have to walk to get round it.

The suitability of a particular hole as a shelter for a bee colony appears to be complex. From examining the properties of places that bees do choose naturally and from putting out artificial 'hives' with different characteristics, we now know that the ideal cavity must (1) have a volume of 15–80 litres, (2) a south-facing entrance, (3) an entrance hole that is smaller than 75 square centimetres and near the bottom of the cavity and (4) be several metres above the ground. Most of these properties are to be found in the beehives that humans put out for their domesticated bees, but, of course, all humans have done is to mimic what a bee would naturally demand in the wild.

Anyway, if after her prolonged inspection, a scout finds a cavity that fulfils some or all of the above criteria, she returns to the swarm still hanging in the tree, and begins to dance about the nest site. The form of the dance is exactly the same as the waggle dance used for communicating sources of food with the direction of the nest site being signalled by the straight 'waggle' part of the figure of eight. The suitability of a site is signalled by the vigour of the dance, with holes that would be an ideal home by all criteria being danced about for up to half and hour and less suitable ones being represented only by sluggish dances.

Up to two dozen scout bees may be 'reporting' back about the different nest sites they have found and so the colony as a whole is getting information about many different potential homes all at the same time. The scouts, however, are sensitive to the dances of other scouts and begin to modify their behaviour accordingly. If one scout has found a hole that fulfils some but not all of the criteria for a good home and dances with moderate vigour and then comes into contact with another bee that has found a 'super site' and is dancing excitedly, the first bee will set off to inspect the second bee's site. If it's as good as it's cracked up to be, she will return to the swarm and dance about it too, switching her allegiance from her own to the new site. Other scouts, constantly coming into contact with bees all dancing vigorously about a particular site will also set out to have a look at it. If it is better than anything they have found for themselves, they, too, will start dancing in its favour. If it is not, they will continue to dance about their own site, causing some of the other bees to go and explore it too.

Gradually, the scouts will, by their repeated visiting and dancing, stop dancing about sites that are deficient in some way and transfer their support to really good ones. The two dozen or so original potential sites are whittled down to a short list of just two or three until ultimately a consensus is reached about which place would be the best new home for the colony, which is the one it goes to. The key to the success of this process is the willingness of scouts to transfer their support from site they have discovered themselves to an even better one discovered by another scout (would that human interactions were as efficient!). The result is that the colony 'chooses' what is probably the best site in the area, at least as assessed by the majority of scout bees. This decision-making may take several days to complete, because each individual bee's assessment of each nest site is so thorough and because it takes time for the 500 or so scout bees to compare the various possibilities and come to their majority decision. All the while the main swarm is camped in a tree and is then, when the decision has been made, guided by the scout bees to the new home. As Thomas Seeley (1977) puts it, 'Thus the nest-site selection process, involving hundreds of scouts, dozens of alternative dwelling places and probably several thousand independent decisions by the scout bees, draws to a close'. A quite extraordinary complex result has been achieved by rules of thumb that are in themselves probably very simple. A scout bee walks around the interior of a nest hole. How long she takes to do this has an effect on how vigorously she dances when gets back to her swarm. How vigorously she dances relative to other scouts affects which sites will be visited by other bees, and so on. The end result is a colony decision about what is the best nest site in the area that is achieved with an efficiency and accuracy to be envied by humans attempting to reach decisions by consensus politics. It also looks, on the face of it, so complex that a 'mind' thinking rationally about the best thing to do must be behind it.

But it is not. Each step could be just an automatic response requiring no thought at all. Simple rules of thumb, assembled by natural selection into activation at the right time (search for flowers as colony is building up; search for small dark holes when swarming; search first and then dance, etc.), are all that is required to explain these examples of bee behaviour. At the very least they should make us extremely wary of assuming that just because animal behaviour looks complex and could be described by using 'mind-laden' terms like 'decision-making' or 'assessment', these are justified as anything more than technical descriptions. Bees, like conjurers, should make us sceptical of what happens in front of our eyes. This, however, raises an even greater difficulty. Are we to say anything that can be explained by a rule of thumb cannot have consciousness behind it? Does understanding how something works (for example, measuring volume by distance walked) abolish 'mind'? Are we only going to allow an animal to have thoughts, concepts and subjective experiences as long as we don't understand what is going on? Having gone through the exercise of chopping animal achievements down to size with Occam's razor (invoking the Clever Hans effect, statistical oversights and rules of thumb to do so), are we then going to conclude that because they are not perhaps quite as clever as we thought, they can't be quite as conscious either?

Again, I ask for the reader's patience. We will eventually have to confront this question of the relationship between complexity of underlying mechanism (in this case 'cleverness') and what might constitute evidence for consciousness. The fact that these questions have raised themselves in such a forceful way at this point only emphasizes how far we still have to go in sorting out the various complexities of our puzzle of consciousness. All we have done so far is to point out that what seems to be complex behaviour may in fact arise from a relatively simple mechanism or rule of thumb, with no necessary suggestion that the animals following them are consciously thinking about what they are doing or even that the mechanism underlying the behaviour is particularly complex. Indeed, we have been looking at examples in which quite simple mechanisms are distinctly plausible. Our next task, then, is to look for evidence of cases where behaviour does seem to result from animals 'thinking' about what they are doing. Even if we find it, however, we must be careful not to take 'consciousness' and 'thinking' as the same thing because something akin to thinking can take place without it necessarily being experienced in the full glare of the conscious mind.

If this sounds strange, think of the many times when some sort of mental activity must have been going on in your brain but you were not conscious of what was going on. Reciting poetry you didn't even know you knew, would be one example, or using the rules of grammar (as shown by correct use of language) without having any idea as to what those rules might be would be another. Although it does indeed sound odd to say so, we have to conclude that although thinking may be one manifestation of consciousness, not everything we think about we do consciously. This means that even if we could demonstrate that animals can think, we would not be all the way to our goal of showing that they were conscious. However, we would have made some progress in that direction. If we could show that they can 'think' and are not just blinding following preset rules of thumb, we would at least be on the right road. The qualifying heats, as it were, would have taken place even though the final race might not yet have been run.

'Thinking' (at least in comparison with 'consciousness') is relatively easy to define. Donald Griffin of Rockefeller University, who has been so influential in getting scientists to face up to the issues of animal consciousness, sees it as a process of attending to internal images or representations of objects and events. For him this means that an animal has some sort of inner representation of the external situation confronting it or that it has memories or anticipations of future situations. Thinking may lead to comparisons between two or more representations and to choices and decisions about what to do next based on some sort of assessment of likely outcomes.

A possible definition of what we mean by thinking, then, is not only that a human or animal should have an internal representation of the world but that it should be able to perform some sort of internal manipulation on that representation – working out what would happen if one element were changed, for instance – and behaving appropriately according to its changed representation. It is the anticipation of the unexpected, the power to be 'one jump ahead' by working things out internally that distinguishes true thought from rules of thumb. Thinking is therefore most likely to occur in animals that have a great many possible courses of action open to them and where, consequently, working out in advance what the best course of action might be is a great deal quicker and safer than trying each one in turn and seeing which one is best in practice. An animal that has a simple list, albeit a long one, of what to do in all circumstances ('if there is a small hole, enter it', 'if there is a sudden shadow, flee', etc.) has no need for recourse to an internal model of the world: it is all prescribed. But an animal that is capable of rising above its preset list and coping with the unexpected, shows that it has the ability to work out for itself which of several possible courses of action will pay off. Appropriate (beneficial) behaviour will be possible even when there are no rules of thumb to tell it what to do. That is why, in our pursuit of an animal 'mind', it is so important to eliminate the possibility of simpler explanations. If all an animal can do is to follow a set of rules, then there is no reason to suspect it has a mind – that it can 'think'. But if, when confronted with novelty, an animal can work out what to do for itself when the task it has been set is complex and free from all the other difficulties that can contaminate its performance, then we may well want to suggest that it can really 'think'.

Just to put your own mind at rest, all the studies of insects that have been done so far suggest that, although they have taken rules of thumb to great lengths, they are not capable of real thought. (I am deliberately including here studies that suggest that bees have a 'map sense' because I do not think that these adequately control for the other possibilities – but that is another story). For example, the female digger wasp that builds a burrow in the sand for her egg and then provisions the burrow with food for the larva, does something very clever but she is clearly a victim of her own rules of thumb.

The wasp digs the burrow and then returns to it several times carrying items of food before she finally seals it up. Each time she returns, she leaves the prey (which is a paralysed but living insect) at the entrance to the burrow, opens up the entrance, goes in to inspect the burrow and then comes back to the surface and drags the new prey down inside.

The great French entomologist Jean Henri Fabre showed how dependent the female wasp was on her simple rules by moving the prey a few centimetres away while the wasp was inside inspecting her burrow. Normally, in the cricket-hunting species he was studying, the female leaves the paralysed cricket on the surface with its antennae just touching the burrow entrance. Emerging from her burrow after Fabre's intervention to find that her prey was not in the position she had left it, the wasp moved it back to where it had been. Instead of then immediately dragging it down into the burrow, however, she left it on the surface and went back down for another inspection visit. Again Fabre moved the cricket and again the wasp moved it back and then went on yet another inspection visit of her burrow. Over 40 times he did this to the same wasp and over 40 times the wasp moved the cricket back – and went on over 40 inspection visits. Not once did she pull the cricket down into the tunnel, not until Fabre stopped interfering and left the cricket alone, whereupon she promptly took it underground. She was following her rules of thumb ('approach burrow', 'leave cricket touching burrow entrance on surface', 'enter burrow', 'exit', 'pull cricket from burrow entrance underground') and was entirely unable to escape from her ritual when, by changing the position of the prey, Fabre had effectively forced her to go back to an earlier stage – the 'approach burrow' stage over and over again. The fact that she had already inspected her burrow (to excess as it happened) did not seem to register at all.

What Fabre and the many other people after him who have done similar experiments with digger wasps did was to reveal the animal's dependence on relatively simple rules by putting the animal into a novel situation and showing that it behaved 'stupidly'. In this case, the element of novelty is relatively minor – a prey being moved a few centimetres – but it was a situation with which the wasp (or probably any of her wasp ancestors) had not had to deal. If a wasp leaves a cricket with its antennae touching the burrow entrance and then goes underground for a few seconds, it is very likely (barring some major catastrophe, such as a herd of cows kicking the cricket away) that it will still be there when she re-emerges. And if a cow has, perchance, kicked it away, then pulling it back and having another look at the burrow will probably be quite appropriate. But a curious entomologist, deliberately making life difficult over and over again is not a contingency for which natural selection has equipped her. The rule of thumb, which normally works very well (how clever of the wasp to make sure that the burrow is still all right before she takes the cricket down!), is exposed as the automatic process it is and the wasp behaves inappropriately, wasting time and effort in inspecting instead of getting on with provisioning her nest.

Rules of thumb do not, of course, have to be innate or built in – they can be learnt. But even learnt rules of thumb do not necessarily imply any very great cleverness on the part of the animals following them. Racoons can be trained to put coins in slot machines, circus animals can be trained to perform all sorts of tricks and a cat locked in a cage can learn to put its paw on the handle and let itself out – all by relatively simple processes of trial and error. To train a racoon to put a coin in a slot machine, you first put some coins near the animal. Then, when by chance its paw happens to touch one of them, you reward it with a bit of food and do so over and over again until it reliably learns to touch the coins every time you give them to it. Next, you stop giving it food, even when it touches the coins, until it has moved one of them with its paw. When it has learnt that trick, you feed it only when it actually picks up a coin, and so on. Gradually, you can train it to do what you want, teaching it one simple step at a time before going on to the next until you have a whole chain of simple learnt responses, which, when taken together, look very impressive and as though the animal is being very clever. But all you would have done would be to instil into the animal a few simple learnt tricks that it had to perform in the right order.

Similarly, the cat in its cage does not necessarily think about how to escape, 'clever' though the way a trained animal escapes may look. The first time it is put there, it wanders round, seemingly at random until by chance it might put its foot onto a spring that opens the door. With a little more practice, it will go straight to the spring every time, having learnt the relatively simple rule that putting its paw gets it what it wants. Only if it could immediately work out what to do if the cage were suddenly changed in some way would be in a position to say that it really thought about what it was doing instead of blindly following the same procedure (albeit one acquired by learning) every time.

So the key to saying that animals are doing more than just following preset rules of thumb and therefore possibly thinking about what they are doing, is to see how they react in *novel* situations – that is, situations which neither their own 'innate' behaviour nor simple learnt rules of thumb could tell them what to do. Natural selection and basic trial-and-error learning can between them achieve a great deal with simple rules, particularly in constant or predictable environments. But in dealing with unpredictability and the challenge of the new (when 'new' means that neither the individual nor its ancestors have met that situation or anything similar before) – that needs something more. That needs 'thinking'.

In the next chapter, we will take on board the lessons we have learnt from this one – that animals can appear to be clever when they are not really, either because they are taking their cues from us or because we have the sorts of brains that make us see them doing clever things because we want to or because rules of thumb can give the superficial appearance of cleverness in the right environment. We will then ask: 'How clever are they really?' 'Can they deal with novelty and work out what to do for themselves?' In other words, can animals think?

In being very fussy and selective about what goes into the next chapter and leaving out many of the examples that people think ought to be included – I do not mean to imply that the evidence for animal thinking is very scanty. The fact that we will not be discussing dolphins or whales, for instance, does not mean that I think these animals are incapable of any sort of mental activity. It just means that the kind of evidence which we have seen from this chapter is essential to say one way or the other is simply not available to us. And the fact that we will not bring up again the examples of 'clever' animals we discussed in Chapter Two, like the vampire bats and the house sparrows, does not mean that they are not genuinely clever after all; it means we cannot yet judge. They may, for all I know, be actively thinking about what they are doing but, as of now, we do not know how they achieve what they do and, until we do know, we shall have to regard them merely as candidates for having real thought not as already having an established watertight case. What we are going to do now is to look at studies of animals that have survived all the onslaughts of Occam's razor that have been wielded at them and still reveal, as far as we can tell, genuine cleverness.

Chapter Four

Thinking ahead

Kingsley spoke slowly.
'As far I am aware, these events can be explained
very simply, on one hypothesis, but I warn you it's
an entirely preposterous hypothesis'.

Fred Hoyle, 'The Black Cloud'

As WE SAW IN THE LAST CHAPTER, THERE ARE TWO BASIC ATTRIBUTES TO what we call 'thinking'. The first is that the thinker should have some sort of internal representation of the world in his, her or its head. This means that it does not just respond to the stimuli immediately surrounding it but carries a memory of things that were there in the past but are now gone or are out of sight. The second is that something is done to that representation to enable the true thinker to work out what would happen under new circumstances – for example, if everything were turned upside down or one element were changed. The classic case is of rat being taught to run a maze and then finding that part of its usual route has been blocked off. Can the animal work out in its head which alternative route it should now take, making use of a representation of the entire maze that it might (or might not) be carrying in its head? If it can go straight to the right route without trying several unsuccessful alternatives first and discovering the way by 'trial and error', then it could be described as having a mental representation of the maze and thinking about which way to go.

This working things out in the head is what we will pursue in this chapter. We will be looking for evidence that animals really can think in this sense; that is, that they can hit upon the clever solution to a problem when neither their natural instincts or their having already learnt the answer beforehand can explain their behaviour. We have already seen that animals will often opt for a simple solution if one exists. If they can get what they want by simply watching their human trainer who has the power to give them food or approval or whatever, then there is no need for them to do anything more complicated than pick up the cues that are ready and waiting and are probably a great deal more reliable than trying to work out the mysteries of some baffling intellectual task. That is not to say, as I have stressed before, that the animal could not perform the task in a genuinely clever way, but if a simpler solution offers itself because some loophole has not been closed, the really clever animal might be expected to take it.

For this reason, most of the examples we are going to look at now are drawn from studies where the animals could not possibly be using humans to give them cues; in other words, where the Clever Hans effect is ruled out by the humans being completely out of the picture – out of sight, out of earshot and preferably not even present as observers. The examples will also be statistically valid, either because large numbers of animals were used or because small numbers were used but were tested repeatedly and there was a clear statement of what would have heppened if pure chance were guiding their behaviour. So it is fairly clear how the first two of the 'simpler explanations' we discussed in the last chapter should be eliminated.

Plugging the leaks against the third possibility – that the animal is following some relatively simple rules of thumb – is, however, rather more difficult because there are often so many different possibilities that have to be ruled out. The experimental design of a study may have to be so complicated to deal with all of them that the descriptions of what was done get a bit convoluted. I hope that these will not put you off. Skip to the conclusions if you wish but I do want to convey the full rigour of what was done in order to justify the conclusions that I draw. So I hope you will stay with me all the way and appreciate the skill with which simpler explanations were systematically excluded.

Let us begin with the most basic ideas of what we mean by thinking. It has to do with 'working things out in the head' and so we need to devise some way of showing that this is what an animal is doing. One of the simplest kinds of working things out in the head is extrapolation – an animal is shown something which then disappears and it has to work out where the object will reappear, given that it is behaving in some predictable manner. So, for example, if a piece of food is being dragged along in a particular direction and then goes behind a screen, will the animal look for the food at the place where it disappeared (no extrapolation) or at the far end of the screen where it is due to reappear (true extrapolation)? An animal that could anticipate where something is due to reappear would be showing a rudimentary ability to work things out in its head. That, at least, was the reasoning behind a study carried out by Julie Neiwork and Mark Rilling at Michigan State University, in which they showed pigeons a clock face with a hand that could disappear and reappear. The point of using the clock was not to try to get the pigeons to think about time but to have a stimulus (a clock hand) with a regular and predictable movement, the position of which could be easily extrapolated when it was made to disappear.

The clock face in the experiment had a single hand which started at 12 noon (0°) and moved at a steady speed of 90 degrees per second. They arranged it that when the hand reached 3 o'clock (90°), it could disappear and reappear again at some other point on the clock face some time later. Their question was whether the pigeons could work out where a hand that had disappeared at 3 o'clock should reappear after a given length of time if it had been moving at a constant speed. Could they, in other words, extrapolate the movement of the hand when they could see it to where it should end up even when they could not see it?

For their experiments, Neiwork and Rilling arranged it that one of three things could happen. In each case, the pigeon saw a clock face with a hand that moved steadily from 12 o'clock to 3 o'clock. Then, either the hand would keep going at the same steady speed until it returned to its original 12 noon position. Or, the hand would disappear from view at 3 o'clock and then reappear at some other point on the clock face, such as 4.30 or 6 o'clock, at exactly the time that would be expected if it had kept moving at its same constant speed in the meantime. The third alternative was that the hand would disappear at 3 o'clock and then reappear at a time and place completely inconsistent with it having moved at the same constant speed – that is, taking too much or too little time to reach 4.30 or 6 o'clock.

The pigeons were provided with a pecking key and trained that they would only get food if they pecked it when they were shown a clock hand moving at a constant speed, regardless of whether it remained visible the whole time or disappeared for a while (in other words, one of the first two possibilities above). If they were shown a hand that disappeared and then reappeared again at an unpredictable time (possibility three), they got no food however much they pecked. Quite surprisingly, the pigeons could learn to do this. It would not have been very surprising to find that they could tell the difference between a clock hand that disappears when it gets to 3 o'clock (case 2 and 3) and one that stays visible right the way round the clock (case 1) but it is quite a feat for them to have recognized that there is a similarity between case 1 and case 2 (the hand moves at a constant speed regardless of whether or not it disappears from view) and that both of these are different from case 3 (hand disappears but must be moving at an unpredictable speed while it is invisible because of where and when it reappears). Nevertheless, that is what the pigeons appeared to be able to do. But there is just a possibility that the pigeons might not have been truly extrapolating and they could just have learnt the specific chracteristics of the clock-hand's behaviour under these three conditions. Perhaps they learnt some rules of thumb such as 'if hand stays visible, peck and get food' (case

1) and 'if hand disappears and reappears at bottom of clock after 2 seconds, peck and get food (case 2) but 'if hand disappears and reappears at bottom after a time that is longer or shorter than 2 seconds, do not peck' (case 3). These three rules would enable them to give the appearance of correctly extrapolating the position of the hand, without them really being able to do so. The critical thing is what happens when the task is changed and the crucial element of novelty is introduced so that none of the old rules will now operate.

Suppose that the hand disappears, as before, at 3 o'clock and then re-emerges at a totally new place on the clock face, one that the pigeons have never seen it appear from before? Can the pigeons then tell the difference between a hand that has been moving at the same consistent speed and one that has been moving unpredictably? The answer is that they can. The pigeons immediately recognized the difference between a clock hand that emerged at a new position (5 o'clock, say, or 7 o'clock) at a time that was consistent with its having continued to travel at 90 degrees per second while it was hidden and one that must have travelled either faster or slower than this to appear at the time and place that it did.

So it was not just the delay itself that the pigeons were noticing but whether the observed delay was consistent with the hand moving at the same speed that it was showing before it disappeared or whether it was not. In order to do this, they would have had to extrapolate the movement of the hand while it was not visible to work out where it should have got to after a given time delay. Having an internal representation of the world and performing some sort of transformation on it – in this case, extrapolating its movement under the assumption of constant speed – is what we have already decided constitutes the rudiments of thinking.

Perhaps, however, 'rudiments' is the right word. Predicting when and where a clock hand should reappear is not exactly a towering intellectual feat even though it does indicate that the birds were showing some ability to work things out internally. The next step forward is to look for evidence of more complex 'thinking' and, for this, we can turn to a study by Herb Terrace, the same Terrace who reared the young chimpanzee Nim and then decided that many of the more startling claims for chimpanzee achievements could not be upheld by the available evidence (although he remained convinced, incidentally, that chimps have intellectual powers considerably greater than our methods have yet revealed). For this study, Terrace turned his attention to a systematic attempt to show that animals (pigeons again) could be taught to carry in their heads the concept or thought of 'order'.

Now to show that an animal is genuinely thinking about the order in which things occur, a large number of simpler explanations have to be ruled out first. It is no good, for instance, simply pointing to the fact that animals often behave in the same set sequences. Male ducks are notorious for the clockwork-like nature of their courting displays – dipping their bills in the water and then raising them skywards, not just always in the same order but with the same split-second timing of the movements as well. But we would certainly not be entitled to conclude that, therefore, male ducks have a concept of order, any more than we would be entitled to conclude that people did if all we had to go on was the observation that they tended to take their handkerchiefs out of their pockets before blowing their noses rather than the other way around.

Nor would we be able to conclude very much about whether a pigeon was thinking about order, simply by training it to peck four objects in the right sequence. This is in fact very easy to do. All you need is a Skinner Box (which is a machine that delivers food if a pigeon pecks in the right place) modified so that instead of the usual single pecking 'key' (really an illuminated disc that the pigeons peck at), there is a row of four keys all of different colours. You teach the pigeon that it is not going to get any food unless all four keys are pecked in the right order. The pigeon learns to peck the red key, then the green one followed by blue and yellow. If these four colours are always presented in the same positions, however, the pigeon could simply be learning to move systematically along the row, from left to right, say, pecking each key it came to. In that case, it would be doing nothing more complicated than learning to move its head from one side to the other, pecking as it does so. There is nothing particularly clever about that, once you know the rule of thumb.

What Terrace did was to train his pigeons in a way that ruled out the possibility that they could get away with using any such simple rule of movement sequences. Instead of keys, he made his pigeons peck at a ground glass screen on which spots of light of different colours could be projected, thus enabling him to change the positions of the lights very easily. His pigeons had to peck their spots of light in the right order (shall we say red–green–blue–yellow) but the position of the lights was systematically varied so that a pigeon might first be presented with RED–GREEN–BLUE–YELLOW (solvable by moving systematically from left to right) but then, on the next time round, be shown GREEN–BLUE–YELLOW–RED and still have to peck them in the order RED–GREEN–BLUE–YELLOW to get any food. Even if the bird cracked that one, the next trial would present it with BLUE–YELLOW–RED–GREEN, and so on. The pigeons would be presented with many such combinations during an hour or so's test. Because the presentation of the lights, the recording of which order the pigeon pecked them in and the giving of food if it did things correctly were done automatically by machine, there was no possibility of humans influencing what the pigeons did. The pigeons were enclosed in a box that blotted out the rest of the room and everything ran automatically without humans being there at all.

Terrace quickly taught his pigeons to peck their coloured lights in the right order, regardless of which positions the four colours appeared in. He used large numbers of birds, his result was highly repeatable but he still wasn't satisfied. It was still possible, he argued, that the pigeons were learning how to solve the problem by learning each pattern of lights separately. If there are four colours, then there are only $4 \times 3 \times 2 = 24$ orders in which these can be presented. Maybe the pigeons saw these as 24 different patterns and learnt 24 different responses to match (BLUE–YELLOW–RED–GREEN means peck middle–R far–R far–L middle–L; GREEN–BLUE–YELLOW–RED means something else and so on). We know that pigeons have remarkable visual memories born, perhaps, of their abilities to find their way home over long distances, and so while it would be a considerable feat of memory for them to learn 24 different rules of thumb it would not be so far beyond what we know they can do that this can be automatically ruled out.

Terrace then did an experiment in which a simple or even a complex memory explanation was ruled out by first training his birds on one task involving pecking things in the right order and then giving them a second different task which could be solved by the pigeons only if they had genuinely developed a concept of order and carried it from the first to a second, novel situation. He was thus making use of what we have already discussed as the essential element of showing thought in an animal – its ability to work out what to do straight away in a new situation for which it has not been specifically prepared. Terrace reduced the number of coloured lights presented at any one time to a pigeon from four to three – green, red and blue – but arranged it that each one could appear in any one of eight positions anywhere in a rectangular space. As before, the pigeons had to learn to peck the lights in the right order (GREEN–RED–BLUE) regardless of which configuration they were shown. 36 pigeons were trained to do this, with the exact configuration they saw being varied with each trial. They had to learn that only if they pecked the three lights in the right order were they given food.

All the birds learnt to do this with an accuracy that was well above chance but, of course, at this stage all that they could have been doing was memorizing a lot of complicated patterns and learning what to do to each one. The birds were then presented with a novel situation. Instead of having to peck three colours in the right order, they were now required to peck one colour and two quite new patterns in the right order. The new patterns were a white horizontal line on a black background and a white diamond also on a white background, neither of which the birds had seen before. But exactly what the right order was was varied for different birds. For half the 36 pigeons, the right solution was to peck in the order LINE–RED LIGHT–DIAMOND or LINE–DIAMOND–BLUE LIGHT; in other words, where the coloured light was one of the three they had seen previously and appeared in the same position in the new sequence as it had in the old. Once again, the physical positions of the three stimuli were changed regularly so the birds could not appear to learn the right sequences by learning their spatial positions.

But the other half were required to learn sequences such as LINE–GREEN LIGHT–DIAMOND or LINE–DIAMOND–RED LIGHT, where the coloured lights were in the 'wrong' position with respect to where they had been previously. All the birds were therefore seeing two unfamiliar and one familiar stimuli. All of them had to peck the keys in the correct sequence, regardless of what configuration they were presented in. But whereas one group had the coloured light in what they had previously learnt was the proper place (green to be pecked first or blue last, just as it had always been), the other group now had it in the wrong place.

Terrace reasoned that if the birds had, during their initial training, really learnt to think about the order in which they had to peck things, then this should carry over into the new task and make it easier for them to solve the new problem if a colour appeared in its rightful place, than if it was in the wrong place. The first group should therefore 'crack' the new problem faster than the second group that were having to unlearn their old ideas about which lights should be in which order before being able to tackle the new one. If, on the other hand, all the pigeons were doing was memorizing patterns of lights and not truly thinking about order at all, then both groups should find the new task equally difficult (or easy) to learn since both were being confronted with new patterns. On the 'thinking' hypothesis, then, there should be a considerable difference between the two groups on the new task, but on the 'learning individual patterns' hypothesis, there should be no such differences.

What he in fact found was a major difference between the two groups. The group with the coloured light in the 'right' place in the sequence found the new task very much easier as measured by the number of trials they needed to get up to the preset standard of having correctly solved the task. They had evidently learnt in the initial part of the experiment that certain lights were supposed to be in certain places (first, second or third in the order of pecking) and had used this information in the second, novel situation. The second group, however, confronted with a situation where lights were in positions in a sequence they were not supposed to be in, were denied this headstart in the novel situation and could not use their previous knowledge to tell them what to do. Thus, during the first part of the experiment, the birds did genuinely appear to have responded not to the particular stimuli that happened to be in front of them at any one time (the row of lights on a given trial) but to an abstract concept, the order in which they had to peck. This they then carried over to the second task. They had learnt that blue was supposed to be 'in the middle' of a sequence of three pecks, regardless of where it was physically and what 'the beginning' or 'the end' were. The birds thus seemed to have an internal representation of their external world that enabled them to work out what to do in a novel situation using a generalized idea of a quite high degree of complexity, in this case the 'order' in which events should occur. It is difficult to avoid the conclusion that they 'understood' the concept of order and that they thought about the sequence in which they pecked.

So we now have evidence not just of simple extrapolation but of an internal representation of the world that seems to deserve the description of a simple 'concept'. But the story does not end here. We also have evidence for even higher degrees of abstraction in animals gathered by people who have, as it were, been willing to go into the lion's den and attempted to see whether animals can think about numbers. The reason why this is such an act of daring will already be patently obvious: many of the claims that have been made for 'counting' animals have been subsequently shown to be bogus, with the animals performing their feats without showing any true understanding of numbers or sums at all. So, obviously, the first requirement for any demonstration of true counting must be that any such 'cueing' is ruled out. The second requirement is a very clear definition of what would be acceptable evidence that an animal can really count or even has any concept of 'number' as we know it.

Hank Davis of the University of Guelph in Canada, who has himself worked on the numerical abilities of a large number of animals, recently suggested that there were at least three levels of 'numeracy' in animals. The first is simply whether an animal can make a relative size judgement – such as 'more than' or 'less than'. A bird or a sheep that joins the bigger of two groups may be able to assess that there are more individuals in one than in the other group, but there is no suggestion that they have literally counted how many members there are in each. The second is a kind of estimation based on a recognition of the pattern that different numbers make. It might, for example, be possible to train a pigeon to discriminate between two playing cards, one showing 5:

$$\begin{matrix} \bullet & & \bullet \\ & \bullet & \\ \bullet & & \bullet \end{matrix}$$

the other 2:

$$\begin{matrix} \bullet & \\ & \bullet \end{matrix}$$

because they look different (the elements make a different pattern) and birds are very good at picking up differences in visual pattern. But, again, this would not really involve an understanding of 'numbers'. So Davis created a third category of numeracy, which he called 'counting', where an animal does show an understanding of what '2' or '5' means over and above just being able to show that 5 is 'bigger than' 2 or that the pattern made by five objects is different from that made by two. Within his 'counting' category, he reserves the term 'true counting' for cases where an animal shows evidence of an understanding of 'number' that holds in many different situations and shows an abstract notion of what a number is independent of the exact situation. We, for example, have no difficulty in applying the same abstract concept of 'three' to three successive buzzes of a bell, three lights showing simultaneously, taking the third turning on the right, three sausages, three economic depressions, three cases of mumps, and so on. Our concept of 'three' is abstract enough that it can apply to all these situations and to new ones that we have not yet encountered. So Davis suggests that we reserve the term 'true counting' for this final, highly abstract form of numerical competence, demonstrated by an ability to transfer from one situation to another, and that we should use a term like 'protocounting' for cases where animals have some concept of what a number means but stop short of being able to see the similarity between 'three' in different situations.

Can animals count, then, in any of Davis's senses of the word? Once again, we fall back on the unlikely (and often unliked) example of the rat and look at an experiment done in Davis's lab by himself and his graduate student, Sheree Ann Bradford. They decided that to have the best chance of seeing whether rats could count, they would have to devise a situation in which any natural counting ability of a rat would be most likely to show up. Rats are nocturnal animals, finding their way around largely by smell, touching and hearing. So, clearly, a visual task, the first sort that a human would think of, would probably not be a good idea. What they did, then, was to devise a system of six tunnels that could be arranged in a row along the side wall of a large box, with all the tunnels perpendicular to the wall and with their entrances all facing in the same direction. A rat would be released into the box from a small starting hatch at one end and move along the row of tunnel entrances. All the tunnels had food at the far end and little swing doors at the end nearest the rat, but five out of the six had a blockage half way down that prevented the rat from getting the food even if it pushed the swing door open. The rat had to learn which one of those six tunnels had the food that it could get to without coming up against a blockage.

There were 12 rats in the experiment altogether. For four of them, the third tunnel along the row of six was the tunnel in which they could get food; for another four, it was the fourth tunnel and for yet another four, the fifth one. Since food was present in all the tunnels, the rats could not simply find 'their' tunnel by smell. Nor could they look down the tunnel entrance and see if it had a blockage because the swing door at the entrance to each one prevented them from seeing what was further down. As the starting hatch was always to the left of the row of tunnels, the rats could, however, solve the problem by 'counting' along the row of tunnels. Davis and Bradford took a great deal of trouble to make sure that this was the only way they could find the right tunnel by eliminating the two other obvious cues – smell and position. A rat trained to go to the third tunnel, for instance, might learn what that tunnel smelt (or looked) like. Perhaps the wood that it was made of smelt slightly different from the others, or perhaps the rat itself would leave a 'smell note' on it or the floor of the box nearby. So the tunnels were regularly changed around, as was the floor covering. On one trial, a 'third-tunnel' rat would find food as usual at the far end of tunnel 3 but, on the next trial, that tunnel would be physically moved and put in the incorrect fifth position in the row with no food at the end of it while a new tunnel would be put in the third tunnel position with, of course, food at the end of it. So any tendency the rats might have had to try and identify particular tunnels by smell or other physical characteristic was thwarted because this would not have reliably enabled them to find their food.

The other thing that happened was that the tunnels were shifted up and down the back wall of the box so that they could either be quite close to the starting hatch or at the far end of the box. Tunnel 1, for instance, could be anything from 7.5 to 79 centimetres from where the rat started, tunnel 2 from 20.5 to 91.5 centimetres, and so on. All six tunnels could be squashed up or spaced out or any combination. This meant that if the rat moved, say, 79 centimetres from the goal box, it could find itself beside any of the six tunnels. Or, it could have passed them all or just be coming to the first one. So it was impossible for a rat to run a set distance, say, 70 steps and then expect to find the right tunnel. It wouldn't necessarily be there. The experimenter sat carefully behind the rat (beyond the start hatch) so as not to influence its behaviour and the whole experiment was videotaped for future reference. A correct response was recorded if a rat attempted to open the door of 'its' tunnel. An incorrect response was one in which the rat pushed or even touched the door of any one of the other five tunnels.

Even with all the precautions that were taken with running

this experiment, the rats still learnt the task very rapidly. In fewer than 100 trials, tunnel-3 rats were going to the third tunnel along, tunnel-4 rats were going to the fourth tunnel and tunnel-5 rats were choosing the fifth tunnel. The tunnel-5 rats were particularly interesting because they adopted a different strategy to the others. They went to the far end of the row of tunnels (that is, to the sixth one) but, without attempting to enter this tunnel, they then went back to tunnel 5, as though 'counting down' from the end. All the others appeared to 'count up' from the beginning.

Davis and Bradford then introduced a slight complication to the procedure. They started putting some of the tunnels on the far wall of the box, so that the rats had to run along the side wall (as they had done previously) to find some of the tunnels and then turn a corner to find the rest. So a tunnel-4 rat might find three tunnels along the side wall as before but not find its own tunnel until it had got to the far end and turned through 90 degrees, the correct tunnel now being much further away and round the corner. Even this did not disrupt the ability of the rats to find the correct position. They still went directly to 'their' tunnel even when it was in a completely new place. It did really seem, then, that the rats were finding the right tunnel by a form of counting (at least between 1 and 4). All the other cues they might conceivably have used to go directly to the unblocked tunnel were, it seems, systematically denied to them and so the fact that they could reliably choose the right number tunnel strongly suggests that they had a concept something akin to 'third tunnel along' or 'fourth tunnel along' that was not to do with its smell or physical location but referred to its ordinal position in a row. By being able to solve this problem, the rats seemed to show an elementary understanding of number – protocounting – to use Davis's word. Perhaps they were also, to borrow Otto Koehler's somewhat haunting phrase, 'thinking unnamed numbers'. They certainly appeared to have an internal representation of their world that went beyond its obvious physical characteristics and into an abstraction of something not so very different from our own concept of number. This internal representation could at least enable them to cope successfully with a novel configuration – when the tunnels were put into new positions – even though we have no evidence from this that the rat's concept of '3' or '4' could carry over into totally new situations as we would demand if we were to call this evidence for true counting. In fact, the evidence for true counting in animals in the strong sense Davis means, is quite sparse. I know no unequivocal evidence for it at all, but there is one example of an animal coming very close. This animal is a quite extraordinary African grey parrot named Alex.

Alex is remarkable not just for the fact that he has a large vocabulary but the fact that he uses English words appropriately to get what he wants. Irene Pepperberg, who trained Alex, realized that one of the reasons that most talking birds don't seem to understand what they are saying is that people teach them to speak in ways that make it impossible for them to make the link between the sound of the word and what it stands for. For instance, how is any parrot supposed to learn that 'Hello!' is a greeting given by people when they first see each other if its main experience of the word is to have it repeated many times by people standing in front of it for minutes or even hours at a time? And how is even the most perceptive parrot supposed to realize that the phrase 'Pretty Polly' refers to the bird itself and is not just the sound that a human being makes, equivalent to a dog barking or a cow mooing?

Pepperberg reasoned that if she could devise a way of training Alex so if he used a particular word it would have important and specific consequences for him (such as the correct word for a piece of food being followed by his being given that food), he might then stand a chance of learning that the word and the result were connected. If he could do this, he might begin to show evidence of understanding the meaning of words or at least using them in the right context. The exciting thing was that not only was Pepperberg successful in teaching Alex to use words correctly, she was also able to go on and ask him questions and get answers from him, some of which, as we shall see, related directly to his understanding of numbers.

Her training method sounds bizarre, but it worked. Alex was

placed on a perch near two human beings who were talking to each other and not to the parrot at all. At this stage, Alex was nothing more than a passive observer. One of the human trainers would hold up an object and say to the other 'What's this?'. The object in question would be something calculated to be of interest to a parrot such as a nut, a cork or a wooden clothes peg (parrots like things they can chew). If the second trainer correctly named the object, she would be given it and praised, watched by the parrot. But if she gave the wrong name, she would be firmly told 'No!' and the object ostentatiously taken away and hidden.

At first, such conversations and the giving or withholding of objects went on entirely between the two human trainers with Alex taking an evident interest in the proceedings. Eventually, however, he started to join in, giving the word (or, initially, an approximation to the right sound) for an object he wanted. If he gave the correct name, he, too, would be praised and given the object. From then on the trainers would direct their conversations to Alex – holding up a series of objects and either giving them to him if he named them correctly or saying 'No!' if he didn't. Alex quickly learnt the names of nine different things – paper, key, wood, peg, a piece of leather, cork, corn, nut and pasta. He could say the right name on over 80 per cent of the occasions when he was shown one of them. He could also cope with variations on the original objects, such as a piece of paper that was of a colour or shape that he had not seen before.

Alex proceeded to show his grasp of human words by spontaneously using them in entirely appropriate ways. It is disconcerting, to say the least, to watch a videotape of an unsuccessful training session in which a parrot refuses to co-operate with his trainer and then see it terminated by the bird itself moving off-camera muttering the words 'I'm going away!' His use of the word 'No!', too, had an uncanny aptness. Just as his trainers had said 'No!' if he had given the wrong word to something they had shown him, so he started saying 'No!' back to them if they gave him something he didn't like. If they gave him a cork, he might drop it and say 'No!' As his training continued, he started to replace his typical parrot speech (inadequately rendered as 'RAAAKK!') with the more and more frequent use of 'No!' to express displeasure or non-co-operation, such as when he was refusing to taking part in any more naming trials. Clearly, talking to the natives in their own language was more effective than screeching at them!

It was, however, when Pepperberg moved on to teaching Alex about colour, shape and number that she opened up the possibility of giving real insights into an animal's mind. She taught Alex the names of three colours – 'rose' (red), 'green' and 'blue' as well as two shapes – 'three-corner' (triangle) and 'four-corner' (square). Alex could say 'four corner paper' if shown a square of paper and asked 'What shape?' and he could say 'green' if shown something green and asked 'What colour?'. He would correctly say 'green' even if the object was a shade of green he had not seen before or was itself something completely unfamiliar to him.

Pepperberg also taught Alex numbers, using the same methods that she had used before to teach him the names and colours of objects. She would put Alex on his perch and then hold a conversation with the other trainer in front of him. 'How many?', one of them would say, holding up five wooden sticks. 'Five wood', the other would reply. 'That's right', she would be told, 'FIVE wood' and be given the sticks. Eventually, Alex might intervene with 'I wood' and then later, after watching many similar conversations, he, too, would say 'Five wood' and then be praised and given the sticks to chew.

In the same way Alex was taught to give the correct response to the question 'How many?' when two, three, four or six pieces of wood or corks were held up to him. If he gave the right answer, he was given the sticks or the corks. If he said the wrong number or called the objects by their wrong name, he was told 'No!' and they would be hidden from him. The corks and sticks used in this experiment varied considerably in appearance because, as Pepperberg puts it in her report, 'of Alex's prior manipulation of these exemplars' (he chewed them to bits, in other words). But the variable physical state of the objects he was being asked about inadvertently made the experiment all the more convincing because it tested his ability to correctly identify how many things there were whatever they looked like. Two intact wooden sticks or two chewed ones or one large one and one nearly demolished one all had to be described as 'Two wood', so any simple pattern recognition of what two things looked like was quite impossible.

Once Alex had learnt to say how many corks or wooden sticks there were, he was then asked about the numbers of other objects such as keys, paper or clothes pegs. These were all things that Alex had seen before and knew the names of but he had never been formally shown collections of them and asked 'How many?'. An experiment was done to see whether he could straight away transfer his knowledge of numbers from corks and sticks to these new things. On 145 tests with the new objects, he gave a completely correct answer nearly 80 per cent of the time, saying both what they were and how many of them were being held up. Most of his incorrect answers consisted of giving the correct name but leaving out the number, such as saying 'Paper' rather than what he should have said which was 'three paper'. After further prompting ('HOW MANY paper?'), he was correct about number in 90 per cent of his 'incorrect' trials. As Hank Davis pointed out in commenting on this reluctance to give the number, it could have been because Alex had no particular interest in getting more than one of anything. What he wanted was a clothes peg to chew and it may not have mattered to him how many he was given as long as it was at least one.

So far, Alex's abilities look impressive – he seemed to be able to put a correct 'number label' on collections of different sorts of objects, even those these varied in appearance, and even to transfer this ability from his initial set of objects (corks and sticks) to different ones (keys, pegs, paper, etc.), but what about some of the other difficulties that could be raised, such as a Clever Hans effect or the fact that only one animal was being used in the whole experiment? What about the statistics of results obtained on one bird? Pepperberg was as careful as she could be to eliminate the Clever Hans effect although she was operating within the constraint that a human being had to be there to present things to Alex. Social interaction and getting what he wanted from people had been the essence of the way Alex had been trained and it could not just be suddenly abandoned like that. Her compromise was to employ special testers to carry out the formal parts of the experiments (the ones where I have given percentages for the results). These were people who had not been involved in Alex's training and were told only the bare minimum about what was going on. They were the ones who actually asked Alex in his tests 'How many?' or 'What colour?', and although they obviously knew what the right answer was, they had no special body language that Alex could have learnt and used to cue him to give the right answer. In addition, all formal sessions were videotaped, so errors of interpretation (or overinterpretation) of what he said at the time could be avoided.

Because there was only one animal in the experiment, Pepperberg also had to be meticulous about the way she arranged for the tests to be conducted. For example, if she had given Alex a whole test session on numbers of objects and given him, say, 100 questions of the 'How many?' sort, Alex could have quickly learnt that one out of only five answers ('Two', 'Three', 'Four', 'Five', 'Six') was correct, at least as far as the first part of the answer was concerned. (He had to add the object name, as in 'Two cork' to be completely correct.) Provided he attached the right name to the objects shown, he could therefore be 'correct' on one in five trials simply by producing a number at random or even saying the same one over and over again. In order to cut down on the chances of this happening, Pepperberg never had test sessions where only one sort of question was asked. Thus, a question like 'What colour?' (correct answer: 'Green paper') would be followed by. 'How many?' (Correct answer: 'Four cork') and then the next one might be 'What shape?' (correct answer: 'three-corner paper'), followed by another number question, and so on. This meant that guesswork on Alex's part was completely ruled out because the large number of possible answers (about colours, objects, shapes, numbers) meant that random or chance behaviour would give a very low level of correct responses. The fact that Alex's degree of correctness was so high (around 80 per cent) even with the constant switching from questions about colours to questions about numbers, shapes and then back again to colours meant that he really had to know the difference between the different sorts of question and was able to distinguish the different colours and numbers accurately. Pepperberg's standards of what counted as a correct answer (right number *and* right object on first utterance) and an incorrect one (anything else, including giving the right answer but not immediately) were also stringent enough to establish that he was indeed doing what she claimed without interference from her or from the vagaries of chance.

At this stage, however, it was not clear exactly what Alex was doing. He was clearly capable of attaching the right colour, shape and number labels to things but quite what his idea of 'number' consisted of was still vague. The objects he appeared to 'count' were very variable in size so that he couldn't just have been equating 'number' with surface area or size. They were also very different in appearance so that it seemed unlikely that he was using an estimation of number based on the pattern the objects made. But it was just still possible that he was doing one of these things (two objects do, after all, usually take up less room than four, and two things do make a different pattern to five). The only way to find out was to try and increase the novelty of the situations in which he had to apply his number labels and find out if he could still correctly apply them even to completely new objects, new arrangements and new combinations.

Pepperberg's next approach was, therefore, to try asking Alex about the numbers of totally novel objects, things that had never been used in any of his experiments before. These were such things as bottles of typewriter fluid, toy cars, washers, thimbles or antacid tablets. Obviously, Alex could not be asked what these were as he had not been taught their names, so all he had to do when asked the question 'How many?' (thimbles, cars, etc.) was to give the correct numerical answer between two and six. Even with these entirely novel objects (actually, they weren't completely novel; they had been placed where Alex could see them for a couple of days so that he wouldn't be alarmed by them, but they were novel in the sense that he had never before been asked any questions about them) even here he answered correctly on 80 per cent of trials. So three bottles of typewriter fluid would get the response 'Three' even though he had never before had to give this answer to these particular objects. Alex had clearly transferred 'Three' from a familiar to an unfamiliar situation, showing that he had an idea of what constituted 'three' that was not tied to specific situations or specific things. This is at least some evidence that he was genuinely thinking about a number.

His next task was even more complicated. Alex was shown a mixed set of objects – say two keys and two corks, making four objects altogether, and again asked 'How many?' Again his answers were astonishingly accurate. On 70 per cent of occasions he correctly said the total number of things present, despite the inherent ambiguity of the question 'How many?' when applied to such a situation. ('How many corks?' or 'How many things altogether?'; I don't think I would be sure which was meant without a little further clarification.)

Pepperberg admits that she does not know whether Alex really does 'count' in the sense that we know it or whether he had a much cruder grasp of what it was to assign numbers to groups of objects. He certainly has the ability to give the right 'number label' in appropriate situations and to transfer it to new ones that he has not specifically encountered before. He has, it would seem, gone considerably further than the rats with their 'checking off' of tunnels in a row but quite what he thinks about is still not clear. Pepperberg herself is quite restrained in what she claims for him and she sums up her experiments as simply providing 'evidence for certain, albeit limited, number-related abilities in the grey parrot'. Number-related abilities, certainly, and those that fall into the category of what we mean by 'thinking' – that is, having an internal representation of the external world. Whether we call it protocounting or protothinking or dispense with such prefixes altogether does not really matter. We do have evidence – real evidence – that some animals at least have the rudiments of what in ourselves we would refer to as 'thinking'. And that evidence, although at the moment limited, points the way to what would count as even better evidence. No longer are we in the realm of speculation with never a hope of demonstrating what we are talking about. On the contrary, it is now possible to pin down two essential elements of thought – internal representation and transformation (change) – and to ask whether or not an animal possesses them by means of experiment. We can do this by predicting what it should do in certain situations on the hypothesis that it really is thinking. The situations that are particularly important in this respect are those containing enough novelty that preset rules are unlikely to give the animal the right answer unless it can 'think'.

We thus set up two rival hypotheses. One, the 'following preset rules' hypothesis, makes predictions about what the animal should do in the new situation, usually involving the animal behaving rather 'stupidly' or in ways that are appropriate to the old rather than the new situation. The other, the 'thinking' hypothesis, makes different predictions. Its predictions involve the animal being able to work out in its head what the correct course of action is and to do something new or 'clever' that is appropriate to the novel situation it finds itself in. Evidence of thought is thus to be found in rival predictions about behaviour. Cleverness shows up in what animals do, provided that the predictions are sufficiently stringent and the animals' behaviour could not be more simply explained in simpler ways. In other words, showing cleverness in animals is dependent on how clever we are in devising ways of showing that their apparent cleverness could not be explained in any other way. We have to hack away with Occam's razor for all we are worth and be as sceptical as we know how ('A bird counting? Ridiculous!'). But when all other possibilities have been exhausted, it may still be the case that the simplest hypothesis that explains all the known facts is the one that says that the animals are thinking about what they are doing. No other explanation for the pigeons' ability to peck three keys in the right order explains what they can do (especially their transfer to a new order problem) as well as the one that says they have a concept of 'order' – that is, an internal representation of what it is for things to occur in sequence. No other explanation explains the rats' ability to find the second, third or fourth tunnel in a row (especially when the tunnels are moved to completely new positions) as well as the one that says they have a rudimentary ability to count. And no other hypothesis explains Alex's behaviour as well as the one that attributes to him the ability to think about numbers. As Sherlock Holmes once remarked: 'How often have I said to you that when you have eliminated the impossible, whatever remains, *however improbable*, must be the truth?' We may be left with preposterous hypotheses, but if they explain the known facts better and more simply than any other, then they should be accepted – at least until yet more evidence shows them to be false. Occam's razor itself demands that we should.

The most plausible explanation, then, for the examples we have looked at so far in this chapter is that at least certain animals can think. It is probably no accident that I have used examples from birds and rats to come to this conclusion because these animals make people immediately sceptical. Tell someone that a chimpanzee or a dog can do something clever and they will be much more likely to believe you than if you tell them that the same thing has been done by a bird. They will demand much more evidence if it is a feathered brain that is supposed to be achieving something than if it is an animal that is more obviously 'like us' in external appearance or behaviour. It is a curious kind of prejudice that makes us assume that the more an animal looks like us or interacts with us, the cleverer it must be, but it is a very common one. The result is that evidence about bird cleverness tends to be the best documented because, to have gained any degree of acceptance, it will already have been poked, scorned and generally pulled to pieces before it ever sees to light of the scientific literature. Only grudgingly will birds or rats be admitted as having thoughts, whereas more 'human-like' animals may be attributed with them on much flimsier evidence.

Now it may have occurred to some readers that much of the evidence we have looked at so far in this chapter is a little, shall we say, artificial. Pigeons do not normally look at clock faces to get their food and most parrots are not, in their everyday lives, required to assess the numbers of bottles of typewriter fluid presented to them. The rats, on the other hand, searching for the nth tunnel in a row, seemed to be doing something that would come naturally to a rat living in the wild – and they solved their particular problem with notable ease. Perhaps, then, in our search for thinking animals, we should venture out of the laboratory (despite all the advantages it holds for controlling the extraneous variables that constantly threaten to leap out and confuse the issue) and look at more natural situations in which wild animals are, without any interference from human beings, required to be clever to survive. There would, of course, be the danger that we could not as convincingly dispose of all the simpler hypotheses that we might want to and there would not be the same ability to manipulate the situation and surprise the animals with the novel circumstances that we have seen are the key to showing that they have true thought. But we might be tapping into abilities that they have that our man-made experiments would never pick up. Just because an animal might come out as not very bright on a human intelligence test or just because an animal finds certain difficulties in mastering human words, does not mean that it would be unable to solve a more natural task for which its particular type of intelligence has equipped it. And, if it is true that the most complicated sorts of task that any animal, human or non-human, has to face are those to do with solving the complexities of its social interactions with other animals, where better to look for evidence of 'thinking' than in social behaviour? Cleverness may come from a successful social life.

There are two potential problems with trying to work out how clever animals really are by looking at their social behaviour in the wild. The first is the one we have already mentioned – the difficulty of controlling everything that has to be controlled and of designing proper experiments in the field. The second is that the most interesting situations – that is, the ones where an animal is confronted with a really novel situation and shows evidence of original thinking about a solution – are going to be single, never-to-be-repeated incidents. In other words, anyone watching these animals is going to be left with nothing but a series of anecdotes that may be fascinating stories about what particular animals did on particular occasions but which by definition can never be seen again. Even if they did occur again, the circumstances would no longer be novel and so the whole point of them would be lost. Science does not like anecdotes. It wants quantitative repeatable measurements, which is often just what cannot be supplied. So what we will do is to look first at one study where a pair of researchers have been able to set up 'proper' experiments with wild animals while still being able to retain the essence of the animals' wildness and natural social behaviour. Then we will look at what might be called 'disciplined anecdotes' – case histories of individual incidents that have been collected together and subjected to a great deal of scrutiny with a view to finding out whether the animals involved understood enough about each other to be able to practise deliberate deception in new and unprecedented situations.

We have already come across one of Dorothy Cheney and Robert Seyfarth's remarkable wild experiments before, in Chapter Two, when we were looking at the way in which vervet monkeys extract a great deal of information from the insignificant-sounding 'grunts' of other monkeys. In that particular case, Cheney and Seyfarth carried out their experiments by hiding tape recorders in the grass at strategic points so that sounds could be played to the monkeys in ways that were not all that different from the way in which they might hear a real monkey that happened to be temporarily out of their sight behind a clump of grass. So they were aiming for minimal interference with the life of the wild monkeys but at the same time managing to be in complete control of the timing and sorts of sounds the monkeys were hearing.

Cheney and Seyfarth used somewhat similar techniques to uncover a whole range of knowledge that the vervets had about each other. One thing that the monkeys apparently knew quite well was how reliable another individual monkey was as a giver of calls, particularly those relating to the presence of other monkeys. Vervets have two calls that they give in the presence of monkeys other than those belonging to their own immediate group – a 'wrr' and a 'chutter'. They live in small groups and get to know each other very well, so the sighting of another group of monkeys is quite an event in their lives. Vervets are also territorial and actively defend their patch against other monkeys. If neighbours do come too near, they are frequently met with the loud trilling sound described by Cheney and Seyfarth as 'wrr!'. This seems both to alert a monkey's own group that there are other monkeys around and also to inform the other group that they have been seen. On hearing the 'wrr' call, the group members come together, often so close that they are touching each other and, together, look hard at the other group. The other group will probably 'wrr' back and be answered in turn by more 'wrr's. Very often, this exchange of 'wrr' calls is all that happens and the groups then go peacefully on their respective ways with no further interaction, although they keep an eye on each other for a while. At other times, however, things get a bit more serious and 'wrr's are followed by active chasing and threatening of the other group, particularly if there is something worth having between them, such as water or a tree with ripe fruit on it. Then there can be skirmishes, chases and even physical contact to settle the dispute. Under these more aggressive circumstances, the monkeys (especially the females who take territorial defence very seriously) will give a second type of call, known as the 'chutter', which is a harsh sound with a much broader range of frequencies than the 'wrr', and also considerably louder. So the monkeys have two calls that sound different but are given in somewhat similar circumstances.

Cheney and Seyfarth decided to make use of the overlapping but slightly different usage of the two calls to discover how much the monkeys knew about the identity and reliability of an individual giving the calls. What they did was to tape-record the 'wrr' call of a particular individual and then play it back through a loudspeaker hidden in the grass to the rest of its group. They played it over and over again when there were no strange monkeys present, so that the 'wrr' call that usually signified the presence of other monkeys became inappropriate and signified nothing. The other vervets would initially look up when they first heard the tape recording, see that there were no other monkeys around and then go back to feeding. Eight times Cheney and Seyfarth would do this to a particular monkey and so by the time the ninth (test) call was played, the monkey in question would probably not look up at all. It had got used (habituated) to the call and realized that it signified nothing. The monkey whose call was being repeatedly played through the loudspeaker was thus being made to artificially 'cry wolf' and come to be seen by the others as an utterly unreliable witness. Now, if the 'wrr' call of that individual had come to be seen as unreliable, would the quite different 'chutter' call of the same individual also be regarded in the same way and also seen as not worth bothering about?

So having habituated some vervets to the 'wrr' call of individual monkey A, Cheney and Seyfarth then looked at their responses to 'chutter' calls recorded from the same individual A. The result was that the other monkeys hardly responded at all, whereas they still continued to respond to both the 'wrr' and 'chutter' calls of other (still apparently reliable) individuals with prolonged and immediate looking around as if searching for the intruding group whose presence was indicated by the call. This means that not only had the monkeys learnt to selectively ignore one call that was supposed to indicate the presence of other monkeys but in fact no longer did when it was given by a particular individual, they had also transferred this ignoring to a different call of the same individual. If animal A had come to regarded as unreliable because one of its calls repeatedly failed to match up to what was really going on, then another of A's calls was also ignored. So it was not the call itself that the other monkeys had learnt to ignore but who was giving it. And it wasn't just the gross acoustic structure of the call they were noticing, because, if they had been doing that, they would have habituated to the 'wrr' calls given by other individuals. As it was, response to other individuals giving the 'wrr' call remained intact but a completely different sound of individual A was also branded as unreliable. All this suggests a quite considerable knowledge of other animals in the group as individuals and an ability to use this knowledge even in the highly artificial situation when one previously reliable individual is suddenly made to appear unreliable by means of a human being interfering with a tape recorder and playing its calls in inappropriate situations.

But, as I said before, the juiciest, most tantalizing evidence for what animals can do by way of thinking about their social situations and working out how to get the best out of them comes from accounts of single incidents in which animals seem spontaneously to hit on the right solution. We have seen enough by now to realize the pitfalls that lie in wait for anyone rash enough to claim, on the basis of a single incident, that an animal is really doing something clever. Like water that will always find a hole if there is one, animals seem always to find the easy way out if one is to be found, leaving us at risk of being overimpressed and gullible. With due caution, therefore, let us for our final examples in this chapter see what those who have studied animals scientifically at close quarters have produced by way of convincing one-off accounts of how animals deal with novel social situations. The most impressive examples come from cases where animals use their social skills to manipulate the behaviour of other animals and so to get what they want by deception.

Deception means that there is an intent to mislead. For animal deception, where, of course, it is not possible to ask the animal what its intentions are, Richard Byrne and Andrew Whiten from the University of St Andrews, have suggested that we should look for four criteria as to what might and what might not count as deception: (1) The behaviour of the deceiving individual should be part of its normal repertoire. The essence of deceit is that it is some thing or some act that is going to be mistaken for something else, so its 'normal' use should be well accepted. Lies are not going to be believed if they are totally outrageous – only if they can be made to seem like plausible day-to-day occurrences. (2) It must be used rarely in its deceitful context. If it is used too often, other animals will come to recognize the deceit for what it is and take no notice. Lying is most successful when it is taken in the context of generally truthful uses. (3) The behaviour must be used in such a way that another individual is likely to misinterpret it (by taking it in its usually honest way) and (4) that the deceiver gains something by its deceit.

For example, in baboons, very obvious gestures of 'looking' are given when one animal sees a predator or another troupe of baboons. Other animals follow the direction of the gaze, usually immediately, so that 'looking' forms a normal part of the baboon's behaviour, and it is important for other animals to respond to it because it may mean danger of some sort. Byrne and Whiten describe several incidents in which this 'looking' was used to manipulate the behaviour of other animals to the lookers' advantage. On 23 May 1983, a young male attacked a younger baboon which screamed, bringing several adults running to the scene. These adults were giving aggressive 'pant grunt' calls, apparently going to attack the young male who, seeing them coming, suddenly and ostentatiously 'looked' into the distance even though there were no predators or other baboons there. The adults stopped and immediately looked in the direction of his stare. Their attack on the young male ceased and he made his escape.

On 8 September 1983, an adult female was feeding, watched by a juvenile. The juvenile approached. The adult female made no threat but carried on eating. The juvenile then screamed. Screaming by a juvenile normally signals that it has been attacked and other adults, particularly a dominant male, will often come along and chase off the attacker. Screaming here also summoned an adult male who promptly chased off the feeding female. The juvenile then took over the food patch and started feeding itself.

Just anecdotes? Yes, of course, but how about this: Hans Kummer, who has studied baboons for many years described an occasion when members of the troupe he was watching were all resting. Then, over a period of about 20 minutes, one of the females gradually shifted her position over about 2 metres so that she ended up behind a rock where she began to groom a subadult male. If the dominant male of the group had seen this, he would have attacked both of them, but from where he was sitting, he could only see the female's tail, back and the top of her head. He could not see her front or hands and he could not see the subadult male, which had bent down behind the rock. In other words, the adult male could see where the female was but he could not see what she was doing.

All these instances could be straightforward cases of learning with nothing very clever or insightful about them. The juvenile baboons could have learnt that the easy way to forestall attack was to 'look' into the distance and the easy way to get a larger animal out of the way was to enlist the aid of another by screaming. The female could have learnt that the only way to get a bit of peace and quiet was to go behind a rock. Initially, they could have discovered what to do by chance – perhaps it really did happen on one occasion that there was a predator present just at the moment of attack and perhaps the young male 'looked' for all the right reasons and then discovered that this had the unexpected result of getting him out of a tricky situation with his social superiors and repeated it subsequently. In this case, he would have been clever enough to learn the connections between events but not really any cleverer than a cat opening a puzzle box or a racoon putting a coin in a slot machine. There would be no justification for saying that the animals worked out in advance what to do or that they had any intention to mislead the 'duped' individual.

But there are two features which cast doubt on this interpretation. The first is that they did not happen all the time. They were rare events, interspersed with 'real' occurrences where no deception was being practised. If an animal has just learnt a trick, then we would expect it to practise it all the time, not to hold back and only do it sometimes. The second is the way the behaviour was performed, particularly in Kummer's example of the female baboon grooming the young male. Kummer himself said he would have been tempted to dismiss this whole incident as purely accidental had it not been for the surreptitious movements the female made to get herself behind the rock. If she had learnt that going behind a rock was a good way to avoid being chased by the male, we would expect her to go straight behind the rock, and not to spend 20 minutes edging slowly towards it. This is what we would expect her to do if she realized that the male was watching her and if she were deliberately trying to deceive him. And her position behind the rock – with the back of her head visible to the male but not her face or hands – was also the ideal position for deceiving another individual. He could see she was still

134

around but he could not see that she was grooming another male. We could explain this as a complicated learning process (she had learnt that she would be chased if she adopted one position with respect to the rock but not another) but the simplest explanation now becomes that she knows what would aggravate the male and was deliberately deceiving him by trying to make her movements towards the rock as inconspicuous as possible and then to hide what she was doing from his view when she got there.

The point is that we could give an explanation of the behaviour based on a learnt rule of thumb that did not postulate any very great social cleverness, but it would be an extremely cumbersome explanation, full of 'she just happened to learn this' and 'she had already learnt that'. Occam's razor now favours genuine cleverness: the young female baboon realized that sudden movements would attract the male's attention and that he would chase her if he saw her grooming another male. If she simply 'worked out in her head' what to do from her social knowledge (and knowledge of what another could see), she could immediately do the right thing, whereas tying it all down to trial and error would suggest that she had spent many long and painful hours gradually learning what to do.

We do not know, of course, exactly what had been going on beforehand so it is strictly speaking impossible to sort out which of these explanations is the right one. It looks very much as though deliberate deception was being practised, with the animals using insights drawn from their previous social knowledge of each other, but, strictly speaking, we do not know how novel the situation was and therefore how much the female had already learnt about the behaviour of other animals. What we need, then, are cases where the previous social interactions between animals are known and where, therefore, we know how new, apparently novel social situations were to them.

Emil Menzel, from the State University of New York at Stony Brook, set up a group of six young chimpanzees in a large paddock and kept a very close watch on the development of their social interactions. He was particularly interested in the possibility that they might be able to use each other's behaviour as clues to where food was hidden. He would begin one of his feeding trials by deliberately showing one animal where food was hidden and then seeing whether the others could infer from its behaviour where it was. All six chimps would be shut in a pen so that they could not see the paddock and six pieces of fruit would be hidden somewhere in it – under leaves or grass or behind a tree, the exact positions being changed for every trial. Then one of the chimps would be carried out in the arms of an experimenter and shown where the food was but not allowed to touch it. It would be put back with the rest of the group and then all six animals would be released together, one of them knowing where the food was and the other five not knowing. On control trials, which alternated with these 'knowledgeable leader' trials, the food was hidden as usual but no animal was shown where it was. This was to control for the possibility that the 'ignorant' chimps could have found it anyway, by smell or some other cue, even when not having help from a leader. In only one of these 46 control trials was the food found at all, but on all 55 trials where one chimp had been shown the whereabouts of food beforehand the food was found by the entire group (within 2–3 minutes in most cases). The 'ignorant' chimps began by following the leader but eventually they followed only if the leader had been shown fruit – a desirable food. If the leader had been shown vegetables – much less sought-after food – the others did not follow nearly as readily, apparently realizing that the hidden food was not going to turn out to be that worth having.

The first evidence of deception began to occur when one of the 'leader' chimps, called Belle, was repeatedly chased from the food (which she had led the group to) by a dominant male called Rock. If Rock was not present in the group, Belle (having been shown where the food was) would lead the others directly to the food and, such were the (reasonably) amicable relationships within the group, that all the other chimps nearly always got something to eat. If Rock were present, however, he would rush over, kick or bite Belle and take all the food for himself. Belle, not surprisingly, started going more and more slowly if Rock were present (while still going directly to it if he weren't there) and soon stopped revealing the whereabouts of food altogether. She would simply go to where it was without uncovering it and sit on it and wait until Rock went away. Rock, however, quickly learnt what she was doing and when she sat down for more than a few seconds, he shoved her off, searched where she had been sitting and got it all.

Belle's next move was not to go all the way but to stop near the food, not on top of it. Rock's response to this was to search the area around her until he found it. Belle began sitting further and further away and would wait until Rock looked in the opposite direction before she moved towards the food. Rock in turn took to looking away and even wandering off until Belle started to move. If ever Rock got close to the food, Belle showed nervous movements that Rock interpreted as 'getting warmer'. On a few occasions, however, she started a trial by leading the group *away* from the food and then, when Rock was occupied with looking for food in the wrong place, she ran over to where it really was and managed to get at least some for herself. In yet another series of experiments, Menzel hid one large pile of food and then, about 3 metres away, an extra piece. Belle then led Rock to the single piece of food and, while he was eating it, rushed to the big pile. Menzel reported 'When Rock started to ignore the single piece of food to keep his watch on Belle, Belle had "temper tantrums"'.

It is clear that Belle and Rock were engaged in an escalating arms race of who could get the better of whom. One would succeed for a while, perhaps only once – and then the other would find ways to outwit him or her. There are two ways in which such arms races could come about. Either Belle and Rock could, when unable to get what they wanted, behave essentially at random, producing some new behaviour that just happened to be appropriate, like the naive cat shut into its cage. Desperate to escape, the cat 'tries anything' and eventually puts its paw on the handle after having put it everywhere else. Perhaps Belle and Rock, each hungry for food, similarly tried out a huge range of behaviour and one of them resulted in giving them what they wanted, even if only temporarily. But the cat, going through many different behaviour patterns, takes a long time to get itself out of the cage for the first time, whereas Belle and Rock did not have this time available. They went straight to the next stage in their war of escalating wits. Sitting on the food rather than eating it was not one of many possible alternatives that Belle tried. She did it straight away. And the one behaviour that would get Rock the food under these circumstances (search where Belle was sitting) was the one behaviour he did. He inferred; he thought about where the food might be; he didn't wait around to develop rules of thumb by trial and error. His behaviour and her response and his response to her response and all the subsequent bluff and counter bluff that went on between them are most plausibly explained on the hypothesis that each understood what the other intended to do. What they did was too quick, and involved going too directly to the correct solution to be easily explained by learning without any insight as to what the world was like. Here were two animals apparently thinking about how to get what they wanted and acting appropriately and cleverly in a social environment that was posing increasingly complex intellectual problems.

Both from field studies and from more carefully controlled laboratory experiments, we have, then, at least some evidence that animals can 'think'. The evidence is not as substantial as we might like but then the study of thought in animals is a relatively recent one. We can at least say, however, that it is no longer an impossibly vague hope that we might one day find better evidence of animal thought than we have now. We can even see what that evidence might be. It will be based on well-designed, carefully controlled experiments in which simple hypotheses (such as that the animals are following set sequences) can be ruled out. It will be statistically valid and, above all, it will look at the animal's response in novel situations, ones in which they could not possibly find the right solution unless they were genuinely 'thinking'. The pigeons pecking spots of light in the right order or 'predicting' the appearance of clock hands show us how this might be done. The rats that go to the numerically correct burrow and Alex with his spoken answers to questions about numbers give us a glimpse of some of the abilities we might find. Even anecdotes about baboons and chimpanzees keep us going for, although not as controlled and repeatable as our more critical selves might ideally demand, they point unerringly to minds that can work out what to do next and can take into account what another animal might do. For the moment, it is difficult to find any other explanation that satisfactorily explains all the findings, at least not with such economy and plausibility.

Thinking is, however, but one facet of 'mind'. The other, considered by many to be the more basic and morally the more important aspect of consciousness, is the capacity for subjective feelings – for experiencing states such as 'anger', 'pain', 'pleasure', and so on. It is the capacity for experiencing, consciously, the states which we know in ourselves as emotions that worries many people about the way animals are treated. Cleverness is one thing; suffering is something else. In order to decide how to treat animals we need to know not just whether they are capable of intellectualizing about their world, but what they *feel*. If we want to gain a full understanding of consciousness in another species, therefore, we must look for evidence of what is, for us, the most compelling aspect of our mental lives, that which gives us heaven or hell, glory or despair, and see whether in other animals, too, there is evidence of an inner life of feelings and emotions.

Chapter Five

Feeling
our way

I imagine that my subjective experiences are very different from yours, except in one particular – like you I regard painful emotions as emotions I wish to avoid and vice versa for happy emotions.

Fred Hoyle, 'The Black Cloud'

OUR EMOTIONS PROVIDE US WITH PERHAPS THE MOST VIVID OF OUR conscious experiences. The way in which a nagging pain can completely preoccupy us or anger temporarily get in the way of our doing or saying anything sensible highlights one of the most important features of consciousness – its attention-grabbing, centre-stage quality. If we are in pain or feel an emotion, there is no doubting that we are experiencing something in the here and now. There is an immediacy to these conscious experiences, of things being currently felt or currently dealt with that makes their reality impossible to deny, although, of course, not all our emotions will necessarily force themselves onto our conscious stage. Ever since Freud, unconscious emotions have been recognized as potentially having very powerful effects on us, driving and affecting our lives even when we are unaware of it, so we must be wary of equating 'having emotions' with necessarily being conscious of them all the time. Nevertheless, as far as consciously experienced emotions are concerned our awareness of them can be intense. Our question for this chapter is whether other animals have them too. Do they know what it is to feel anger, fear or a longing for something that is not present? Are they strangers to happiness? Do they know the joys of parenthood or the pain of loss?

In attempting to answer such questions, we are going to need rather different methods from those that we used to study the intellectual capacities of animals in the last chapter. There, as we saw, the kinds of internal processing that were probably going on inside the heads of animals were deduced from their behaviour in essentially the same way that one might deduce the workings of a computer program – by putting them through various tests that can be passed only if they possess a certain degree of intellectual sophistication. So, if we want to know whether an animal can 'count', we set up a situation in which only an animal that can really count could get the right answer and where one without this ability would get the wrong answer. But with emotions, no such attainment tests seem possible. There are no right or wrong answers for emotions so they do not lend themselves nearly so readily to benchmark criteria that can be compared across humans, computers and any other species of animal. What we need to look for is some way in which emotions, if they were present in other species of animal, would be most likely to reveal themselves. We are after the outward and visible signs of inner emotional experience.

In ourselves, these outward and visible signs are bodily changes such as blushing and rapid heartbeat as well as behaviour such as clenching of the fists, smiling, laughing, and so on. Note that these observable signs are not the same thing as conscious experience of the emotions themselves. Laughing is not the experience of happiness, nor are tears sadness itself. The conscious experience of these emotions remains private and could be suppressed or even be faked by an actor or a spy, but we are usually reasonable confident that we can use what is visible to an outside observer to indicate what is invisible and going on within.

There is an obvious problem in relying on outward signs of

inner awareness when we go beyond the boundary of our own species. The expression of emotion that we see in other people and can usually interpret in them at least moderately well, may be quite different in other animals. A bear covered in fur is not going to give very much away if and when it blushes, and birds, by the constraints of their anatomy, cannot raise a clenched fist or even smile. Even if we were to acknowledge that each species has its own unique ways for giving public expression to its emotions and we were prepared to spend a lot of time finding out what they are, we would still have difficulties making the link between emotion and behaviour. What looks like 'smiling' in a chimpanzee, for instance, and could easily be interpreted by us as a sign of happiness, turns out to be given in the presence of aggressive, dominant animals when attack is likely and so is most probably a sign of fear. To identify reliably emotions in other animals, therefore, we would probably do well to look beyond particular, species-bound behaviour that may make it extremely difficult for one species to understand another and turn to more general ideas of what emotions are and how they find expression.

For us, one of the most crucial aspects of feeling an emotion is that it *matters* to us. If we have a pain, it matters to us that it should go away, or if we have an ambition, it matters to us that we should achieve it. We are not passive bystanders to our own emotions, calmly noting down life as it passes us by. We are, whether we like it or not, involved. Our emotions have caught us up and made us care, often passionately, about outcomes. They have made us into beings to whom results (getting away from danger, having a drink after a long period with no water, being accepted into a group) matter. This means that if we want to know whether other animals have emotions, what we have to look for first are signs that they, too, care about what happens to them and that for them, too, it matters what the world is and does.

At this point, we would appear to be terminally frustrated by language, or rather the lack of it, in other species. If I want to know what matters to you, I can ask you. You could deliver a passionate speech about the things that were important to you – family, career, collecting old silverware or whatever it happened to be – and I could, within quite a short time, have a very clear picture of the things that mattered to you. I obviously cannot do the same with any other animal. I can sit and watch them and listen to the sounds they make but on the subject of what they care about, if anything, they remain obstinately silent.

But perhaps we would be wrong to insist on speech to tell us what we want to know. Perhaps by doing so we would be letting our human obsession with language get in the way of an even more fruitful source of information. After all, although language is a rapid and sophisticated way of discovering what matters to other people, it is certainly not the only way we have of doing so, even with people who speak our own language. In some ways, that people *do* is an even better guide to their emotional state than what they tell us in so many words that they are feeling. Someone who says on one occasion 'Oh! I wish I could play the piano!' is going to be a lot less convincing about what they really want than someone who says nothing but simply gives up 4 hours every day of their lives to practise. Someone who 'puts his money where his mouth is' will convince us of the importance he attaches to a particular project far more than someone who simply talks about how much he values improving the environment, bringing jobs to the unemployed or whatever it is without doing anything. We are actually quite rude about people who just use words and do nothing else. 'Windbag!', 'A lot of hot air!', 'Mere words!', we say. It always rather amuses me that people elevate the human species above the rest of the animal kingdom by virtue of its use of language and then denigrate talk that is not followed up by getting on with the job. Language may be precious but words are often regarded as cheap.

Anyway, what matters is that language is far from being the only way we have of finding out what matters to other people and at least some of the other ways – for instance, the actions that convince us that they mean what they say or really feel strongly about something – are also open to us when we ask about other species. The absence of a convenient buffalo, bowerbird or baboon language in which we could ask them in words what they feel is, therefore, certainly not the formidable obstacle it might seem at first. On the contrary, the fact that we are forced to use more direct and compelling methods with them may actually be an advantage. At least it is more difficult to lie if you cannot use words.

We have so far used rather casual examples of people expressing what matters to them without the use of words – practising the piano every day or giving sums of money to an inner-city regeneration project. To translate this idea of using actions in this way into a form that could be applied to non-human animals, we now have to do two things. We have, first, to be more explicit about how an action can be used to show what matters to the person doing it – that is, to say what it is about somebody's action that makes us believe that they have strong convictions or strong emotions. Not everything they do will lead us to conclude that they are feeling strongly and so we must be able to pick out whatever it is about some of what they do that makes us think that on this occasion they do have strong feelings. Then, second, we have to think of animal equivalents of these human actions. Animals don't have money to give or pianos to practise on, so if we want to use their actions to tell us what they are feeling, we are going to have to be somewhat imaginative devising situations in which they can reasonably tell us what we want to know.

These two questions are closely bound up together, so let us begin by trying to be more precise about the human situation. An extremely rich man who gave a small amount of money to, say, a school in a deprived area would certainly be making a welcome gesture, but he would only be making a small contribution relative to all the money he had. He would be making a gift but, in terms of the difference it would make to his own lifestyle or even his accumulated capital, it would not cost him very much. On the other hand, a much poorer man, who gave exactly the same amount, might literally have to dig much deeper into his pockets to do the same thing. For him, giving to the school might constitute a genuine choice: if he gave the money to the school, he might then be unable to pay for repairs to his own house.

Taking these two men, both of whom might have given exactly the same amount of money, we would have to conclude that it was the poorer man – the one who would have had to give up something of importance to himself to give the money – who was showing the strength of his feelings for the school. The school would clearly matter to him enough that he was prepared to suffer considerable personal discomfort (such as a leaking roof to his house, let alone not being able to go on holiday that year) to make sure it survived. So it is not just what is given or done that leads to the conclusion that he felt strongly about what he was doing, but what he had to give up to achieve it; in other words, what it cost him. Only if the richer man had given so much to the school that he, too, started to feel the pinch and his own standard of living began to fall as a result would we start to think that the school mattered as much to him. The story of the widow's mite – the widow who gave the smallest possible coin but it was all she had – comes down to us as the quintessential example of the small, but costly, gift.

Money is a relatively straightforward way in which people pay costs because with a fixed and limited income, paying for one thing (a cruise on a yacht, say) often does mean that money for other things (food, house repairs, video-recorders, etc.) is simply not going to be available. The choices that people make under such circumstances give a clear indication of the priorities they put on various things – how much each of them matters to them, in other words. Money is not, however, the only way in which people pay costs. The young pianist who voluntarily gives up 4 hours a day all through her teenage years when all her friends are out enjoying themselves is demonstrating how much music matters to her by virtue of what she has *not* been able to do with the time spent at home practising. The limit of 24 hours in a day forces each of us to make choices of how we spend our time, and what we do – including what we forgo – gives a clear indication of what matters to us. People who uproot themselves from their homes and flee to another country where they may be materially much worse off demonstrate that it matters to them that they go. People who camp all night on a cold pavement to get tickets for a tennis tournament show how much being able to see a critical match matters to them. (Watching tennis may be trivial to us, but they are demonstrating by their behaviour just how different their priorities are from ours.) 'My kingdom for a horse!' shows what really mattered in the middle of a medieval battle. In each case, what counts is not just what is done or obtained, but what is given up to get it. 'Mattering' is not just a question of an action but the cost of an action in terms of time, money, possessions or even, sometimes, a career, a marriage, a reputation, a life or just leisure.

This huge diversity in what humans regard as costs and in what currency they are willing to pay to get what they want means that our second problem – how to devise ways of asking non-human animals what matters to them – is beginning to solve itself already. If humans can express what matters to them in such a variety of ways, then other animals can be put into much the same situations. Apart from those about money, we can ask them exactly the same sorts of questions-without-language that we can ask about other humans: are they prepared to make an effort, spend time, give up an opportunity to feed, etc. to get what they want? An animal prepared to 'do anything' to escape from a small cage or to cross an electrified grid to get at a female is showing what matters to it just as much as humans with cheque books and shaky bank balances.

In theory, then, we can 'ask' animals what matters to them, using actions not words. The clearest answers so far have come from experiments done by psychologists training animals to perform actions for various sorts of reward. We have already seen the results of several such experiments, particularly the ones in the last chapter when pigeons had to peck keys or rats had to run down corridors to get pieces of food. There our interset lay in seeing how clever the animals were at working out how to get a reward. Now, by turning our attention to the value the animal places on the reward itself, particularly in relation to the effort it has to make to get it, we can learn not just what, but also how much, something matters to it. For instance, when a pigeon has learnt to peck a key for food, will it still keep pecking when instead of having to give just one or two pecks

147

per item of food, it has to peck four, eight or even 50 times? Does the pigeon regard a couple of grains of wheat worth the effort of pecking so many times at an unresponsive key? We can ask even more interesting questions by changing the nature of the reward, say, by giving a male pigeon a brief sight of a female if he pecks the right key instead of food. (This can be quite easily automated using a door that slides up to reveal a female pigeon behind it.) Male pigeons will quite readily learn to do this even when no food is involved, often performing courtship behaviour when the female appears. We can then ask whether the male will still keep pecking even when he has to work harder and harder (more and more pecks) for just one brief sight of the female. We can even ask how the male regards the sight of a female in relation to food by seeing whether he is prepared to work as hard for a female as he is for food. All sorts of questions about how animals view different aspects of their environments can be framed by seeing whether they will work (in this case, peck more and more times) for what they want. We see what matters to them by arranging that they have to make a real effort to get what they want. They are then free to pay the price or abandon the attempt altogether.

The beauty of this method is that it can be adapted to almost any species and used to investigate almost any aspect of its environment. Animals with hands like those of rats or racoons are easy to work with because they can be given levers or little handles to press and large birds can be given keys to peck. Smaller birds or animals less able to manipulate things can be trained to hop onto a special perch or sit on a panel built into the floor. Even animals with no beaks or paws, like fish, can be trained to swim through hoops or tunnels. As long as there is some device, like a photobeam, to pick up where the fish is, then the animal can be automatically rewarded for swimming through one hoop rather than another or even just for going to one corner of its tank. And what counts as the 'reward' for the animal learning these various tasks can be varied too. Sliding doors can reveal mates, rivals or offspring and the reward can be an opportunity to interact with them in some way. Or the reward could be a piece of nest material or a chance to hear a recording of the call of its own species, or indeed anything else that the experimenter thinks might be valuable to the animal and is ingenious enough to think of a way of presenting in small doses. Sometimes the things that animals regard as rewarding and are evidently prepared to learn quite complex tasks to have are quite bizarre to our way of thinking. Butterfly fish, which are brightly coloured fish found extensively on coral reefs, find it rewarding to be able to see a very crude model of a smaller fish, which, in nature, removes parasites from their skin by eating them. The real parasite-removing fish is a wrasse that, because of its role in life, is known as a cleaner fish. When a captive butterfly fish is shown a model of its cleaner – which does not, of course, remove any parasites from it – it rubs up against it and shows the special 'invitation to be cleaned' posture that it would show to a real cleaner fish. George Losey and Lynn Margules found that butterfly fish would learn to go to a particular place in their tanks simply to be shown a model of their cleaner fish, even though there was no food there and no parasites were removed from them. Nevertheless, they would go there again and again just for the sight of the useless model.

However strange and however far removed from our own ideas of what is worth having, therefore, we can find out what a particular animal wants and how much it is prepared to pay to get it if we ask it in the right way. If an animal will repeatedly go to considerable lengths to make something appear (or conversely to get it to disappear), then it is telling us by its behaviour that it values that commodity (or its absence). It is telling us something about what it wants or finds desirable. Then, if we make it more difficult for the animal to get what it wants by making it pay a higher and higher cost, we can find out what really matters to it. Without the need for any words at all, we find out what an animal's priorities are and whether the animal behaves as though some things are so important to it that it will do anything to get them.

In us, one of the characteristics of feeling a strong emotion is the way it takes over our whole mind. Having a 'burning desire' and 'being in the grip of an emotion' are two ways we have of describing how strongly we feel about something and we translate what we feel into behaviour by being willing to pay a high cost or even 'do anything' to get what we want. So while what we ourselves feel is the emotion, what other people see are our actions, in which we may go all out for one thing, possibly sacrificing a great deal else on the way. The counterparts of those actions can also be observed in other animals, if we are ingenious enough with our apparatus and harness enough tricks of modern electronics (photobeams, pecking keys, etc.) to overcome their lack of human hands and human speech. What might look like nothing more than a few trivial tricks on the part of an animal is, then, potentially an extraordinarily powerful way of asking them what they value and what really matters to them. It is like a window opening onto their worlds, a window that we might have supposed could only be opened through language, but which it turns out can be flung wide open by the simple means of discovering what they will learn to do, and what they will learn to do it for.

There are few more compelling examples of animals telling us what matters to them thatn that which came from a purely accidental discovery by A.P. Silverman, who was at the time working on mice and hamsters. He was in fact doing an experiment that was completely unrelated to asking animals about anything at all. He was interested in the effects of long periods of inhaling cigarette smoke and his experiment was in no way set up to allow his animals to express their 'opinions' of what he was doing to them. Nevertheless, his experiment had to be stopped prematurely because the mice and hamsters found their own way of telling him.

To look at the long-term effects of smoke, Silverman set up a glass container for each animal and arranged it that a steady stream of tobacco smoke was blown into the container down a glass tube. The animals lived in these containers all the time and were provided with food and water but not much else. After a while, and long before any proper results had been obtained, it became apparent that the whole experiment was going to be a complete failure because so many animals had learnt to stop the stream of tobacco smoke coming into their containers by bunging up the ends of the glass tubes with their own faeces. The faeces of mice and hamsters are small, rather neat, oval-shaped objects and are just the right size for this job. The animals learnt to shove several of them firmly into the ends of the glass tubes, thus blocking the smoke. Several of them actually asphyxiated themselves before it was realized what was happening because the smokey stream of air was also their only supply of oxygen. This, however, only served to make their 'message' all the clearer: to have a stream of tobacco smoke (or, as it turned out, any jet of air) pouring down on them continuously was so obnoxious that they would go to any lengths to stop it, even if it meant cutting off their own air supply. They were 'telling' the experimenter, even one who had not set out to ask them, that they regarded what he was doing to them as something to be avoided by any means available. Even if they had had the power of speech, they could hardly have put it any more clearly.

If we can learn so much about what matters to an animal from a chance, serendipitous experiment like this one in which the animals happened quite literally to take matters into their own hands, imagine how much more might we be able to learn from experiments specifically designed to canvass their views. One of the first deliberate attempts to do this was prompted by the recommendations of the Brambell Committee, a British Government body set up in the 1960s to look at the welfare of animals kept in intensive farms. This committee was a pioneering and far-sighted group of people whose recommendations were so revolutionary at the time that many of them have still not been implemented more than 25 years later. Because this committee had such an impact then and has contributed so much to improving the conditions of farm animals in the years following the publication of its report, it is perhaps a little unfair of me to single out one of their recommendations as being subsequently 'contradicted' by the very animals they were trying to help. But, nevertheless, it does illustrate how even well-meaning humans may not always know what matters to the animals themselves.

Among the many things that the Brambell Committee recommended was that if egg-laying hens are kept in cages, the floors of these cages should not be made out of fine wire mesh (chicken wire) but out of stout strands of heavy-gauge wire. Their grounds for saying this was that they thought that it would be more comfortable for the hens' feet if they had something thick and substantial to stand on rather than the finer mesh that might press into their feet. Following the publication of the Committee's report, Barry Hughes and Arthur Black, working at what was then called the Poultry Research Centre in Edinburgh, decided to find out what the hens themselves preferred to stand on. They gave hens a choice between the different sorts of floors discussed in the Brambell Committee report and then looked to see which ones they chose. They allowed the hens to wander freely between the floors and then noted down, every 10 seconds, where they were standing. To nearly everyone's surprise, the hens seemed to prefer the wire mesh (condemned by the Committee) to the thick wire favoured by it. They spent much more time standing on what had been thought to be uncomfortable wire mesh than the heavy-gauge wire. The reason for this became apparent from photographs taken from underneath the hens' cages, looking up at their feet. With the wire mesh, each foot could be seen to be supported by many different strands, and so even though each individual piece of wire was very thin, the foot was supported at many different points. With the thick wire, however, the smaller number of strands meant that each one supported a much greater proportion of the hens' weight. It was like the difference between someone standing on your foot with a snowshoe or with a stiletto heel: if a given weight is distributed across many points of contact, it is far more bearable than if it is concentrated on just one point.

This example, although it illustrates the dangers of jumping to conclusions about what other animals like and dislike, does not really give a fair picture of the hens' view of their environment. It does not, for instance, indicate what they think of being in battery cages in the first place, because a choice between two sorts of wire floors could be said to be a choice between two evils. It is possible that the hens found neither sort of wire particularly comfortable –

any more than it is desirable to have anyone standing on your foot, in snowshoes or not. To understand what hens really feel about their environment, we need to know *how much* it matters to them what sort of floor they have. If it is a matter of indifference to them what is underfoot, then we would hardly wish to say that they have strong feelings about what sort of floor they stand on. But if it matters a great deal to them that they should have, say, something loose to scratch in, then we might wish to conclude that strong feelings were involved if they found themselves permanently on the wrong floor. We might even wish to say that they felt so strongly about it that they experienced serious deprivation or suffered if kept in wire-floored cages. So we need to devise a way in which hens can be made to pay costs and then show whether they are prepared to pay those costs for the opportunity to be on the right floor.

If hens are offered a genuine choice – that is, they have the opportunity to choose not just an artificial wire floor that humans might force them to stand on in battery cages but also a more natural floor where they can scratch and dustbathe – then even the previously favoured wire-mesh floor is quickly abandoned in favour of floors consisting of peat, earth or wood-shavings. Junglefowl, which are the wild ancestors of our domesticated chickens, spend long hours scratching away at the covering of leaves that hides one of their favourite foods – the minute seeds of bamboo. An ancestral memory of this way of life seems to have carried down the generations into the cages of our modern intensive farms so that even highly domesticated breeds have the same drive to scratch away to get their food – if they have the opportunity. If hens that have been kept all their lives on wire floors with no sight or contact with anything that could be scratched or raked over are suddenly, at the age of 4 months, given access to a floor of wood-shavings or peat, even these naive hens have an immediate and strong preference for these more natural floors over the wire ones, which is all they have known until then. They dustbathe, eat particles of peat and scratch with their feet. It is not just the extra comfort afforded by a soft floor that attracts them, but all the behaviour they can do there as well.

This natural behaviour is so important to them that even when they cannot do it 'properly' (with real soil or wood shavings), they will still pretend that they can. Chickens in battery cages which have wire floors and no loose substrate for the birds to scratch and dustbathe in, can often be seen to go through all the motions of having a dustbath. They squat down, raise their feathers, rub themselves against the floor and flick imaginary dust onto their backs. They behave as though real dust were being moved through their feathers, but there is nothing really there. If such dust-deprived birds are eventually given access to something in which they can have a real dustbath, like woodshavings or peat, they go in for a complete orgy of dustbathing. They do it over and over again, apparently making up for lost time when they were unable to do the real thing.

So on all these criteria – what the hens will choose when given the option, their insistence on dustbathing even when no proper materials are present and the way they indulge themselves when finally allowed to – all suggest that having the right substrate underfoot is important to them. They seem to give dustbathing (and all the scratching and feeding behaviour that hens do when confronted with loose material) high priority. This conclusion is confirmed when they are only allowed access to dustbathing and scratching material if they pay a cost to get to it.

There are many ways in which it would be possible to get hens to pay costs for what they want. It would be easy to arrange, for instance, that they had to peck a key before being allowed access to food, leaf litter, nestboxes, other hens, fresh grass or anything else we might want to offer them, but it is even easier and even closer to the situation in which a wild chicken or ancestral junglefowl might find itself, to insert a barrier with a small gap between the hen and what it is thought she might want. The cost to the hen can be varied by varying the size of the gap. A very wide gap – much wider than the hen – imposes little or no cost: the hen has simply to walk to what she wants. But a narrow gap – only passable if the hen squeezes her body and struggles to get through – is a considerable cost. We can show that the hen regards it as such by putting food on one side of the fence and a hungry hen, which can see the food, on the other. Under these circumstances, a hen hesitates. She takes a long time to go through, although she will go eventually if she is hungry enough. It is clear, therefore, that although she regards it as a cost, it is one that she is sometimes prepared to pay.

We can use this reluctance to go through gaps to find out how important it is to hens to gain access to different sorts of floors. Norma Bubier of Oxford University presented hens with various sorts of floors – wire, litter, grass – and arranged it that sometimes they could have what they wanted without any cost (wide gaps) and sometimes they had to 'pay' by squeezing through a narrow (9-centimetre) gap. To give a comparison with other things that hens might find desirable, she also gave them the opportunity to go into a nestbox, sit on a perch, feed or be near other hens, also with and without having to make the extra effort of squeezing through a barrier. Only if it mattered to them to get litter to scratch in or a nestbox to lay their eggs in would they have to pay anything to get them. Bubier found that the hens would repeatedly squeeze through small gaps to get to floors made of loose material, to get at food and to enter a nestbox. Their response to nestboxes was particularly striking during the short period each day when they were just about to lay an egg. They would search frantically for a nestbox, suspending all other behaviour to do so and were very willing to pass through the barrier to get to one. The high priority given to finding a reasonable place to lay suggests that, at least once a day, the millions of hens that are confined to cages without nestboxes experience a strong state of frustration at not being able to find one.

Just to emphasize, in a slightly different way, the point that scratching material and nestboxes are very important to hens, we can look at the results that Bubier obtained on the way hens responded to other hens. Hens are social animals and on the whole choose to be near other birds, particularly ones that they know, rather than on their own. If given a choice between being close to their flock-mates or alone, they will generally choose to join a flock or at least get reasonably close to it. But when Bubier arranged it that they could be near other birds only by pushing through the same-sized gap that they were prepared to negotiate to get to litter or nestboxes, she found that they chose not to do so. The cost they were evidently willing to pay to get somewhere to scratch in was not one they were willing to pay to interact with other birds. This is quite remarkable because much of a hen's life is spent interacting with other birds and so we might have thought that their social life was important to them. Wild junglefowl and small groups of free-range hens tend to move around together and to have highly developed 'peck orders' within their flocks. Nevertheless, it seems to matter more to hens that they can dustbathe, feed in a natural manner and have access to a nestbox when they need it than that they can be social.

We can see, then, that by the judicious use of costs tailored to the particular animal we are studying, we can ask sensible questions about what really matters to that animal. I have used examples so far from laying hens because these are animals I happen to be reasonably familiar with but the same general approach can be used with any animal we like. Take two other animals that could hardly be more different: pigs and Siamese fighting fish.

Pigs are ideal animals for the sort of study we have been discussing because they are intelligent and, once the problem of building apparatus strong enough to cope with their enthusiastic attempts to get food out of it all the time has been solved, co-operate admirably in the experiments. The best way is to provide them with a stout metal panel (the equivalent of a pigeon's pecking key) which they can then push with their noses. They tend to crash their heads against it, excitedly grunting and squealing, and then hurl themselves onto the food that gets delivered when they make a correct response, so that a pig training session has the sound of a major disaster or someone being thrown violently around and screaming in acute agony. Provided the equipment is robust enough, however, pigs quickly learn that pushing the panel gives them food, turns on a light, opens a door to a bed of straw or whatever it is that the experimenter has decided constitutes the 'reward' in that particular experiment.

Lesley Matthews and Jan Ladewig at Trenthorst in Germany

decided to find out how pigs regarded the opportunity for contact
with another pig relative to their desire for food. Pigs, like chickens,
are very social and, in one experiment, easily learnt to push a panel
for the reward of being allowed nose-to-nose contact with another
(familiar) pig. Every time they pushed the panel, a door went up
between the two pigs so that they could sniff each other. In another
set of experiments, the pigs were given a small food reward and, not
surprisingly, they also learnt to push the panel for food.

Matthews and Ladewig then ran two sets of experiments. In
one, the pigs got 20 seconds of social contact for each panel push,
whereas, in the other, they got 27 grams of concentrated pig food.
In both sets of experiments, the 'work' the pigs had to do to get their
rewards was gradually increased from one panel push for one portion
of food (or social contact), then to 2, 5, 10, 15, 20 and eventually 30
panel pushes. This means that by the end of the experiment the 'cost'
of a reward was 30 times what it was at the beginning. As far as food
was concerned, the pigs kept working (that is, paying the cost of
many panel pushes for just one reward) with very little change in
the number of food rewards they obtained. They adjusted to the
increased effort they had to make by paying whatever cost they had
to. Food was evidently important enough to them that they would
work to get it almost regardless of the cost. Their attitude to social
interactions was, however, very different. Although they would readily
push the panel for a 20-second interaction with another pig if all they
had to do was to push the panel once, they would stop doing it once
the cost got too high. Nosing another pig was 'worth' it for a small
effort but evidently not worth it if the panel had to be pushed several
times to get any result at all. In these particular circumstances, it
clearly mattered more to the pigs to eat than to be social.

A similarly higher priority for feeding over a social interaction
is also shown by male Siamese fighting fish, which have been bred
for their aggressive qualities and might, therefore, be thought to have
an overriding concern for being aggressive. It is certainly true that
they find fighting, or at least the opportunity to display to another
fish, rewarding. Jerry Hogan and his animal behaviour group at the

University of Toronto trained fighting fish to swim up a short tunnel for the sole reward of seeing a mirror reflecting themselves for 20 seconds. There was no food at the end of the tunnel and no real rival fish either, but the males mistook the mirror image of themselves for another fish and 'threatened' it by raising their gill covers and extending their fins just as they would to a real fish. The mirror fish (which was, of course, behaving aggressively) provoked them to be even more aggressive and for the full 20 seconds they would be having a mock fight with themselves until the lights were switched off and the mirror was no longer illuminated.

The male fish found it so rewarding to threaten the mirror that they would repeatedly swim down the tunnel to do so. However, when the cost of seeing the mirror was increased so that they had to swim down the tunnel twice, three times or up to six times for just one 20-second view of the mirror, they did it much less. The more effort they had to make, the less they were prepared to do it. When six tunnel swims were called for to give just one sight of the rival, they hardly ever seemed prepared to pay this cost to see the mirror. On the other hand, when the fish were rewarded with food at the end of the tunnel, even repeated six-tunnel swims did not deter them. They worked for as many food rewards when food was cheap (a single swim down the tunnel) as when it was very costly (up to six tunnel swims for just one food item). On these grounds, food seemed to matter more to them than fighting.

It will be obvious by now that the potential for 'asking' animals what is important is enormous and limited only by our own imagination as to how to put the 'questions' correctly. Simple responses, such as animals moving away from some situations or towards others, gives us the first clue as to what they like or dislike. But the techniques that involve putting up the 'cost' of different commodities enable us to go a lot further and to see how important different things are to animals. Animals do not just choose or express a preference. They can be made to put a price on what they choose and to tell us how important something is relative to the other things in their lives. Food provides a useful yardstick in this context because the need for food is to a greater or lesser extent a universal part of the survival of all animals. An animal that tells us by its behaviour gives something such high priority that it is prepared to give up everything else for it, including the chance to feed, is effectively saying a great deal about the importance it attaches to that particular thing and the extent to which its life is dominated by the need for it. At the beginning of this chapter, we saw that one of the characteristics of feeling 'emotions' was precisely this sense of being dominated by the need for one thing, of somehow being driven to get something we want or get away from something we dislike. Emotional states are not just bodily states in which behaviour is likely to occur. They are states in which the entire resources of the body are directed, if only temporarily, to one end, and in which the attention of the mind is focused specifically on that one goal.

We have now seen that in other animals, too, their behaviour and their decision-making processes become dominated by particular needs, even at the costs of having to make a considerable effort or neglect other things that they might be doing. We have, then, shown that quite a major part of what we mean by emotions in ourselves also appear to be present in other animals. For us, there are 'positive' emotions such as those associated with eating when hungry, finding a place of safety and so on, and all of these result in our wanting to repeat the actions that led us to experience them. We also have negative emotions associated with such situations as lack of food, loss of social companions and injury, and these on the whole result in our wanting to avoid the situations that led to them. Other animals, too, experience situations that they will learn to re-invoke or cause to be repeated just as there are others that they will apparently go to any lengths to avoid repeating. So in this, too, we see parallels that are too striking to be lightly dismissed. But is this enough? Is it enough to define emotions in terms of readiness to approach or avoid or even to the extent to which approaching or avoiding dominates the decision-making processes of the brain? Haven't we still left out the most important aspect of emotions, the conscious awareness of pain, pleasure, suffering, loneliness, joy and all the other experiences we could describe as emotional? Aren't emotions more than just what we *do* but overridingly to do with what we *feel* about what we do? In asking whether other animals have emotions, we may have established that in some ways they do as we do but we have not yet answered the critical question of whether they feel as we feel.

It will not surprise you to learn that the answer to this question still eludes us. We can study the behaviour and physiology of other animals as much as we want, but we cannot know for certain whether their obsession with getting food is accompanied by the unpleasant experience we call 'hunger' or whether their fierce defence of their young carries with it the feelings of love and wish to protect that similar behaviour would in ourselves. As we have seen all along, we have a similar problem of not knowing for certain what other people feel and yet that usually does not stop us from empathizing with them and assuming that when they behave in certain ways, they do feel pretty much as we do when we do the same things. In other words, we make do with the argument from analogy to carry us over the logically uncrossable gap between their private worlds and ours. And because we seem to understand and predict what other people will do, the leap of analogy seems quite justified and not such a terrible sin as the knowledge that we have defied logic would make us think we had committed. It seems quite a normal, sensible thing to do, for all it is supposed to be impossible.

With human beings, the 'leaps' of analogy turn out to be a series of little steps rather than a flying jump into the outer darkness of the mind of an unknown creature. We are so like other humans in some ways that our so-called leaps of analogy can seem like nothing more than walking along a path and stepping from one paving stone to another. We cross a tiny logical discontinuity with no effort at all and it seems absurd to insist that we can have no knowledge of what other people feel or experience.

Our near certainty that we can draw more or less correct inferences about the feelings of other people on the basis of their behaviour comes, of course, from the similarity of other human beings to ourselves – in appearance and what they do. With other species, the similarity is less apparent, often much less so, and the tiny step of analogy that we can often take with humans turns out to require, for non-humans, a veritable rocket launch to an unknown planet. At least, that is what we might think. A very remarkable series of experiments by a Canadian physiologist, Michel Cabanac, has, however, reduced the leaps of analogy that we might think we had to make to much more manageable proportions. In fact, it could almost be said that he has, for certain feelings and emotions, brought them almost down to the size of those we make with other people. Perhaps the paving stones are still a little further apart than they are with our own species, perhaps more like crossing a river on a series of stepping stones than the smooth progression from slab to slab we can use with our own species, but the path forward is clearly laid out and needs only a small effort on our part to travel down it.

Michel Cabanac sees a close connection between physiology, behaviour and conscious feelings on functional grounds. We experience feelings of hunger, he argues, because that is part of our mechanism for rectifying a food deficit and getting something to eat. We experience fear and pain because these are part of our body's way of removing us from situations that are life-threatening. There is no question for him: conscious experiences are there as survival aids. His experiments show that the physiological and behavioural parts of the mechanisms in ourselves and those in other species are so similar that the leap of analogy we would have to make to assume that conscious experiences are also similar is reduced to the barest minimum.

One of his experiments involved studying the responses of rats and humans to the taste of sweet substances. For the human part of the experiment, he made up a sugary solution of known concentration and then got people to taste it and to report on what they were subjectively experiencing. He asked them to be quite precise in how they described what they experienced and give a numerical score to how pleasant or unpleasant they found the drink. If they found it very pleasant, they were to give a score of $+2$, if pleasant, they were to rate it $+1$. If it was very unpleasant it was supposed to be given a score of -2, unpleasant -1 and if it was neither pleasant nor unpleasant they were to describe it as neutral and give it a score of 0. Using these numerical scores, Cabanac plotted a graph of changes in peoples' subjective ranking of what they were drinking under different conditions, such as when they had just had a meal or another sweet drink beforehand. Not surprisingly, the pleasantness rating of the sugar solution went down dramatically under these circumstances and then climbed again later. In women, the reported pleasantness of sweet-tasting things also varied with the phase of the menstrual cycle they were in, with sweet things being liked more at some stages than at others.

Then he turned to other animals, in his case rats, which also like drinking sugary liquids. Unable to ask rats to report on their subjective experiences when they were drinking, he nevertheless derived an equivalent numerical score based on how great a volume of liquid they were prepared to drink under different circumstances. He found that the numerical scores based on how much the rats drank were almost identical to the numerical scores on subjective 'pleasantness' reported by the humans. So a person who had just had a meal would say that they didn't think the sweet drink was very nice and a rat that had just had a meal would refuse to drink very much of it. A human who had not eaten for a long time would report that, subjectively, that drink tasted lovely (+2) and a rat similarly deprived of food would drink a great deal, and so on. Plotting two graphs – one of the human reports of their conscious experiences while drinking and another of the amount of liquid drunk by the rats – produced two lines that were so identical in shape that, if you didn't know, you would have thought they had been measured in exactly the same way.

This is powerful justification for drawing an analogy between humans and rats and saying that it is not just their behaviour that is similar but their conscious experiences too. We assume (making the 'easy' analogy between ourselves and other humans) that if someone reports a drink as 'very pleasant', he or she is probably experiencing, consciously, what we do when we say something is pleasant. The strong parallels between humans and rats in their behaviour and physiological changes after a meal makes the more 'risky' analogy – between ourselves and rats – hardly more outlandish. The ways in which the bodies of rats and those of humans respond to lack of food have been shown to be so similar in so many studies that rats are often used as a model of humans in, say, the study of obesity. It is, of course, logically possible that the rat model could work (in the sense of giving exactly parallel results to the human body) in all ways *except* in having conscious experiences. The rat could be a little machine with no feelings and our bodies could also be machines that work in the same ways, but it just so happens that we have conscious experiences and rats do not. This is possible but distinctly less plausible than the alternative view that rats have not just their physiology and behaviour in common with us when it comes to eating but have conscious experiences associated with satisfying their hunger too.

It is not just in taste sensations that humans and rats seem so similar. Cabanac has also done experiments on the way humans and rats respond to changes in temperature and found very close similarities here too. The conscious experiences that human beings reports, such as calling a temperature of 20°C 'very pleasant' or 'very unpleasant' depending on whether we are too hot or too cold, finds its exact parallel in the behaviour of rats given levers that give them heat or cold. If their brains are warm, they cool their skins and if their brains are cooled, they work to warm their bodies. We say that a temperature is unpleasant if it is very different from our normal body temperature and rats with levers start taking steps to lower or raise ambient if it moves very far away from theirs, and the more it moves, the harder they work.

Again, the simplest hypothesis – the one that needs the least special pleading – is the one that holds that because humans and rats are similar in what we can observe (verbal reports or behaviour), they are also similar in respects we cannot observe (conscious experiences). Although in the last analysis, then, we have to fall back on some version of the argument from analogy to conclude that other species have conscious experiences, it is nevertheless a very restricted version of the argument from analogy, a far cry from the idea that other animals are exactly 'like us' in all ways. This bare minimum use of analogy says nothing more than that other species share with us a basic dichotomy in conscious awareness of emotional states. They share a knowledge of unpleasant emotional states which we might, for ourselves, subdivide into pain, frustration, longing, grief and a host of other negative emotions but which we might also describe simply as 'unpleasant', 'to be avoided at all costs' or even 'suffering'. And they also share with us a knowledge of happier states of mind which we might distinguish in ourselves as 'pleasure', 'joy', 'fulfilment' but which, for the sake of the present argument, do not need such a rigid classification.

All that matters is that it does seem plausible that other animals seem to be aware of states of mind that are broadly 'pleasurable' or 'to be gained or regained at all costs'. Each species may have its own subdivisions of the pleasant–unpleasant axis, some of which may correspond to human ones, others of which may be quite alien to us. Discovering in detail what those might be is for the future. For the present, we need assume only that something like our experience of pain/suffering on the one hand and pleasure on the other can also plausibly be said to occur in at least some other species. We know they have likes and dislikes; we know some things can become so important to them that the pursuit or escape from them can dominate their lives and cause theme to pay heavy costs to achieve their goals. And we know from Cabanac's evidence, that the parallels between them and us are, at least in some circumstances, so close that it seems quite churlish to deny that the behavioural and physiological side of emotion occurs without there being some glimmer of conscious awareness of what is going on – a stab of pain or a rush of pleasure – as well.

It seems we have almost achieved our goal of understanding of what goes on in the minds of other species. With the evidence we have accumulated in this chapter on emotions and in the last one on thinking in other animals, we are probably as ready as we can be for the final approach to the citadel of consciousness itself.

Chapter Six

Balance
of evidence

'Of course it's unlikely, but it might happen that
way all the same.'
'Come off it, Dave! What do you predict? Where
d'you put your money?'

Fred Hoyle, 'The Black Cloud'

ONE OF THE THINGS I HAVE TRIED TO DO THROUGHOUT THIS BOOK IS TO
make a clear distinction between the private sensations we refer to
as 'conscious' on the one hand and those that are open to public
scrutiny, such as what an animal can be seen to do, on the other. I
hope I have made it clear that even processes like 'thinking' or
'counting', although they may be invisible, are nevertheless in the
second, public category because their *results* are public. Thus, if we
say that an animal is 'extrapolating the position of an object', we may
be unable to observe the actual process of extrapolating but we can
make predictions about what the animal should do if it is genuinely
performing this operation. When a moving object has disappeared,
for instance, the animal should be able to go to the place where the
object is due to reappear rather than the place where it disappeared.
It is the behaviour of the animal that tells us what has been going
on in its head. Similarly, with a claim that an animal can count, we

167

can make various predictions about what it should do in different circumstances, such as that a 'counting' horse should be able to tap the ground with his hoof the right number of times even if the person with him does not know the correct answer to the problem he has been set. It was the wrong number of hoof stamps (a very publicly observable event) that told us that Clever Hans was not really counting, not a minute examination of the private recesses of his 'mind'. Equally, it was what Alex the parrot *said* (open to all to hear) when shown different numbers of unusual objects that led to the conclusion that he had some concept of number. Drawing conclusions about what is going on inside the heads of animals in this way is no more mysterious and no less scientific than what physicists do when they draw conclusions about elementary particles that they have never seen on the basis of predictions of what should happen under different circumstances. They may not be able to observe the particles but they can observe their effects.

With consciousness, as we have already seen, the situation appears to be quite different. The reason that so many people insist that we can 'never know' whether another animal (or another person) is conscious is that there are no critical predictions that we can make. Every prediction that we can think of – that people who consciously experience pain should cry out if we hit them with hammers, for instance – can be matched by a sort of shadow prediction – that people should cry out when they are hit with hammers without actually *feeling* anything. Our predictions seem quite unable to escape from their shadows and the eternal taunt that there is nothing we can do to distinguish 'really' feeling from 'behaving as if' feeling. This is also the reason why the study of consciousness is often thought to be unscientific. Science thrives on predictions and a theory that makes no predictions can consequently be dismissed as unscientific.

I want now to ask whether the existence of conscious awareness is really as untestable as it is usually made out to be. In other words, I want to question whether the distinction between private (supposedly untestable) consciousness and public (testable) behaviour that we have kept to so far, is as hard and fast as it seems. Then having argued that it is not, I will then be in a better position to draw conclusions about what is the central concern of this book – the existence and significance of animal consciousness.

For many scientists, there are two kinds of questions: those that we can hope to find answers to (everything from the origin of the Universe to how kidneys work) and those that we can never hope to answer. Consciousness is almost alone of all the phenomena of our world in being placed in the second category. There are many questions in science to which we do not, as yet, know the answers, but most of them we can see our way to answering one day, perhaps when our techniques or data storage methods improve. It is in principle possible to discover how an embryo develops from a single cell or what gravity is, even if the full answers are still a long way off. But consciousness seems to be even in principle inaccessible. We can't even see how we would ever discover whether another organism were conscious. The barrier here seems to be a logical one, not a technical or intellectual failing. This, at least, is the orthodox view. I would now like to explain why I think it is mistaken.

A theme running through this book has been that conscious awareness evolved because it was, in various ways, an advantage to its possessors. It is possible that its only possessors are humans and it is found in no other animals except ourselves, but even if that were the case I would still, as a biologist, stick to the view that humans in some way gained, in evolutionary terms, from their possession of this attribute. I am certainly not alone in this view. Many people from zoologists and psychologists to philosophers have talked about the 'functions' of consciousness and speculated on the advantages it might confer. The important point is that if consciousness has a function, it must also have an *effect*. And if it has an effect, that effect is, in principle, detectable. Let me use an analogy to make this clear.

Suppose you met someone who claimed that he had invented a wonderful gadget whose function was to enhance the performance of motor cars. Some cars had it and others didn't but you could never know which ones were which because the gadget was completely undetectable. You could not see it, hear it, touch it and it took up no space. Moreover, he said to you, there was no way in which you could tell whether a car had one because there was no effect on performance either – no increase in acceleration, no economy of fuel, or anything else. Up to this point, you might have been impressed. We are so used to 'mysterious' modern inventions that do all sorts of things that the invisible black box might not have struck you as anything particularly out of the ordinary. But when he says that it doesn't have any observable effect on performance either, then you should begin to think there is something wrong.

If he wants to claim that his invention has the function of enhancing performance, then there must be some aspect of perform- ance (even if it is one that you have not yet been able to measure) that gets enhanced. If his little box does nothing at all, he cannot legitimately claim that it has any function. Either it must be functionless or it must be possible to detect something that it does to some aspect of the car, even if it is very difficult to detect, such as preserving engine life by a few minutes. It must be one or the other. It can't have a function and no effect. And there must be an effect if it is to be justifiably described as having a function.

In the case of consciousness, we have to be very careful that we do not make the same mistake as the enthusiastic inventor with his completely ineffective machine. We can avoid his mistake in one of two ways. Either we can say that consciousness is undetectable because it has no function and no effect on anything whatsoever. We thus become what are known a 'epiphenomenalists' and see consciousness as just 'sitting above' observable events (the prefix 'epi' meaning on or above), basically as something that arises as a separate but non-intervening phenomenon in certain circumstances (for example, whenever a brain reaches a certain level of complexity, or just in humans or whatever your particular view). Or, we can see consciousness as in some way interfering in behaviour, in being part of the mechanism of its control. We see it helping in some way, perhaps by making actions more decisive or in giving better anticipation of the future. We see it as having both an evolutionary function and an effect that we could detect, some way in which an animal possessing it would do observably different things from an animal not possessing it.

Both of these views are logically consistent. What is not consistent is the hybrid view that consciousness has a function but we will never be able to detect it. We must therefore either abandon all talk of the function and evolution of consciousness and become strict epiphenomenalists, or we must face up to the fact that by talking about the function of consciousness, we have committed ourselves to the view that it should in principle be possible to decide whether an animal was conscious by what it did. This might be very difficult to do and in our present state of knowledge we might not know what to look for, but if consciousness can be said to have a function, then it must make animals do something distinctive that we could detect.

As yet, there is no definite way of distinguishing between these two views but the accumulating evidence does suggest that it is no accident that some of our actions are conscious and others unconscious. Bernard Baars, in a recent book called *A Cognitive Theory of Consciousness*, makes the startling but, on reflection, plausible, claim that most of the things we do, we do better if we are not conscious of what we are doing. Anyone who has had the experience of trying to give a musical performance in public will know the disasters that occur when the stress of the situation makes them consciously ask themselves 'What is my left hand supposed to be doing next?'. A piece of music that can be played effortlessly when the performer is relaxed and able to move their fingers unconsciously through complex and well-known sequences becomes a nightmare of wrong notes and missed chords when the conscious mind tries to take over. Unconscious processors are frequently much better than conscious ones at dealing with what is well known and predictable. Conscious processors are better at dealing with novel situations, unpredictability or any situation where something has to be worked out afresh. The very fact that we can identify cases where consciousness is a help and where it is a hindrance suggests that it has a definite function in some circumstances and is not just an 'extra' coming along for the ride.

But if consciousness has a definite functional role, it must have effects on behaviour that we might one day hope to understand. Although we have, at present, no clear idea as to what those effects might be, we do have some strong indications of where they are likely to be found. Novelty, unpredictability and trying to be one jump ahead are the features that seem, in us, to provoke conscious events. They are also characteristic of an extremely important part of the lives of many animals: their social interactions with each other.

We have seen over and over again that animals respond to each other in highly complex ways. From the vampire bats that feed other bats that have fed them in the past to the battle between the chimpanzees Belle and Rock at the end of Chapter Four, we have seen how subtle and discriminating their social lives can be. It was quite deliberate that the discussion on animal thinking ended with what happened when one animal seemed to be intent on getting food and the other seemed equally intent on keeping it from him, because it is precisely in this escalating arms race of social interaction that the major resources of intellect are called upon. One interpretation of what was going on was that Belle did not want Rock to get her food and that he knew how she normally behaved when she went to get it, so she knew she had to do something different from usual to make him think she had no intention of going to where food was hidden. It was not enough for her to do just anything different – she had to do something that would cause Rock to think she was not going to uncover food. The kind of plot and counter plot that Belle and Rock were playing out over food has been described as 'Machiavellian intelligence' by Richard Byrne and Andrew Whiten. It appears to demand a knowledge not just of what an opponent is likely to do in a present set of circumstances, but also foresight as to what he *might* do if circumstances were changed. It might even involve working out what the opponent would do if *he* thought that *you* thought he was likely to behave in a particular way, and so on to even greater degrees of complexity.

172

In 1976, Cambridge psychologist Nicholas Humphrey published a very influential paper called 'The social function of intellect', in which he made exactly this point. For most species, he argued, it is not the physical world that demands particularly great intelligence, but the social world. Even for early humans, finding food and shelter were relatively simple tasks compared to that of surviving in a group of other humans, in which working out who you could trust and who was likely to deceive would become more and more complex as they tried to work out whether they could trust you and what they could get away with if you thought you could trust them, and so on.

This social intellect theory (which Byrne and Whiten, with an admirably developed sense of historical attribution, refer to as the Chance–Mead–Jolly–Kummer–Humphrey hypothesis after the many people who have contributed to the same idea) points to a very important possible function for consciousness. Being able to put yourself in someone else's position to such an extent that you can predict and even manipulate what they do needs a very special sort of intelligence and one which, the social intellect theory proposes, a conscious awareness of the sort 'this is what I would do in the same situation' would be a huge advantage. The function of consciousness would thus be to enable the organism possessing it to work out what to do in constantly changing social situations where complexity was continually being compounded by counterintelligence in its allies and opponents, all using their consciousness to enable them, in turn, to work out what 'self' was likely to do. As we have seen, humans have no monopoly on behaviour that can be described as 'trust', 'deceit' or 'reciprocation'. Because of this Humphrey and others have argued that consciousness is most likely to be found in animals exhibiting a high degree of social cleverness.

But social cleverness is not the only situation in which 'self' may form part of the picture and it may not be the only area in which being conscious gives some benefit. All sorts of other tasks, such as damming rivers, catching fish or retrieving objects from near-inaccessible places, may demand a knowledge of what the 'self' has done or might do and the effects this would have on subsequent events. Philip Johnson-Laird, from Cambridge, drawing on analogies between brains and computer programs, argues that consciousness might be involved whenever the brain effectively follows a self-referring instruction – for example, getting out of the way of an object that its own body has thrown. For him, consciousness is a superefficient programming device, used when 'self' is both director and object. Keith Oatley, from Glasgow, on the other hand, worried that 'the more we understand mental processes, the less need there seems to be for consciousness', believes that we should also consider yet more radical uses for consciousness, such as restructuring the whole way our brains interpret the world when we change our minds or adjust to changing circumstances in our lives. We need consciousness to cope with major upheavals of our world view.

If any of these ideas have even some truth in them, then it follows that there are not two sorts of questions that we have to ask – that is, questions about animal bodies (to which we can one day hope to have an answer) and questions about animal minds (to which we never can). On the contrary, questions about animal consciousness should be brought firmly into the framework of biology as they are as much part of the subject as the study of oxygen-carrying molecules or of feet. Such questions about animal consciousness may be peculiarly difficult to answer and they may demand new ways of thinking, but then questions about the origin of the Universe or the nature of matter have demanded that physicists also adopt new ways of thinking. For all that, they are no less part of physics than the more traditional approaches of Newtonian physics. Biologists may similarly have to re-order their ideas about consciousness or about the way the brain works in order to accommodate new ideas.

Indeed, the philosopher Daniel Dennett argues that the whole way in which we think about consciousness is in need of a radical shake-up and, in rethinking the way we see consciousness, he believes that we should see it as a proper object for scientific explanation. In a book with the uncompromising title of *Consciousness Explained*, he attacks the idea of consciousness as a single 'stream' and even more the widely held view of it as a person inside our head looking out at the world trying to make sense of it (because that just raises the problem of who the person is who is looking out and what is supposed to be going on inside their head, and so on *ad infinitum*). Instead, he sees it as a sort of creative pandemonium, in which many different parts of the brain are all simultaneously doing their various jobs and giving their own version of events. There is no one right version, no one place 'where the buck stops', just fragments continually thrown up by a multiplicity of channels. The illusion of there being a central, unified 'I' is just that – an illusion. I shall not even attempt here to do justice to Dennett's ideas. I shall only advise you to read some of his books. They are wonderfully entertaining as well as being extremely exciting (a bit like being in an intellectual pillowfight!). The point I want to make is that he, too, strongly supports the idea of taking the mystery – although not the awe – out of consciousness by studying it scientifically, even if we have to turn our ideas of what it is upside down in order to do it. Which, being Dennett, he is not afraid to do.

If we are, then, prepared to rethink what we mean by consciousness and what 'it' is, then it may be that we may not always have to think of consciousness as being completely outside the understanding of science or inevitably different in kind from all other biological phenomena. If 'it' or 'they' have function or functions that we can identify, then it or they should certainly be studied as part of science. Indeed without them, our understanding of animal or human behaviour will remain woefully incomplete. The fact that we find it so difficult to tackle the issues raised by conscious awareness in ourselves, let alone in other animals, may be simply a reflection of our current ignorance or the inadequate ways we have of describing it. But 'do not know now' is different from 'will never know'. One says that it is just a question of time before we do understand a great deal more than we do now about animal awareness, the other that there are logical barriers to our ever knowing. If consciousness has an important and serious function in the lives of animals, then it properly belongs in biology and, when biology is ready to receive it, it will take up its rightful place there.

For the present, however, let us use the evidence for complex mental processes in animals (thinking and feeling) that has been presented in this book, and see where it has got us in the search for animal consciousness. We have seen that some animals behave in ways that are best explained on the hypothesis that they have an internal mental world of their own and manipulate it vicariously by 'thinking' in ways that are at least partly like the ways we do it. Concepts of order and of number do not seem confined to the human skull. Outwitting opponents by double-guessing them and using detailed knowledge of the reliability of different individuals are not uniquely human skills. We have also seen that some animals care sufficiently about the situations they are in that they will go all out to change them or, in other cases, to prolong or repeat them, again with parallels to what we do in comparable situations when we consciously 'feel' strongly about something. Now there is nothing that logically compels us to believe that because at least some other species have some similarities to us in these ways, that they are therefore like us in being consciously aware. We could maintain a persistent scepticism and say that similarities of behaviour are achieved with no conscious states of mind whatever. Animals might count, and remember and puzzle out what to do in a new situation, all completely unconsciously. And they could wriggle and squeal and work hard to get away from something with no flicker of real conscious feelings. Logic says that that is all perfectly possible. But logic also says two other things: first, that on the same grounds we would have to allow that other people may not be conscious either, and, second, that some rather special pleading is going to be needed to maintain that similarities in behaviour co-exist with a lack of similarity in conscious awareness.

What this means is that if we accept the argument from analogy to infer consciousness in other people on the grounds that they are like us in certain key ways, then it is going to be very difficult to maintain that consciousness should not be attributed to other species if they have at least some of those same key features. What makes us reasonably certain that our fellow human beings are conscious is not confined to what they look like or how they live or even whether we can understand their language. Our near-certainty about shared experiences is based, amongst other things, on a mixture of the complexity of their behaviour, their ability to 'think' intelligently and on their being able to demonstrate to us that they have a point of view in which what happens to them *matters* to them. We now know that these three attributes – complexity, thinking and minding about the world – are also present in other species. The conclusion that they, too, are consciously aware is therefore compelling. The balance of evidence (using Occam's razor to cut us down to the simplest hypothesis) is that they are and it seems positively unscientific to deny it.

To someone outside science, it may seem surprising that anyone should even try to deny it, but scepticism and even downright opposition to the idea are still common in the scientific community. Donald Griffin, who has perhaps done more than anyone else to make scientists take the issue of animal consciousness seriously through books such as *The Question of Animal Awareness* and *Animal Thinking*, still finds objections among behavioural scientists to the whole idea of studying animal consciousness. He finds that they frequently dismiss it as so much idle speculation or, worse, a taboo subject that no respectable scientist would tackle. These attitudes seem to come from a deep-rooted belief about what 'proper' science is.

In the early part of the twentieth century, an American psychologist, John Watson, published an extremely influential critique of virtually the whole of psychology as it had been practised up to that time. He claimed that most of it, from Freudian analysis to the study of subjective sensations, was worthless because it was concerned with mental states (like thoughts or emotions) which were subjective, undemonstrable and therefore could not be a proper part of scientific enquiry. The only things he thought should be allowed into science were behavioural or physiological events that could be seen and measured by everybody. What went on in the mind, even the mind of another person, was private and unmeasurable and therefore not a legitimate subject for study by anyone who called themselves a scientist.

Behaviourism, as Watson's creed came to be known, maintained a firm grip over many scientists' way of thinking about both human and animal behaviour until as recently as the 1970s. The study of what went on in the minds of animals was kept so far out of the picture that even 'thinking' or 'counting' (which, as we have seen, have externally observable consequences) was looked upon as not really 'proper' science. Since then there have been considerable changes in what people consider to be testable and therefore scientific, so that thinking and even feeling have found their place in the scientific literature. But consciousness is still to many people beyond the pale of scientific respectability and the old reluctance to consider it as a biological phenomenon still remains. Many scientists demand more definite evidence for consciousness in animals before they will commit themselves or they take refuge in the 'we can never know' arguments we have already discussed. In some ways, this scepticism is valuable. We have seen over and over again that studies of mental abilities in animals are peculiarly susceptible to overinterpretation. Peoples' fondness for the animals they work with, the intimate communication that may be going on between them and their animals and a general human gullibility (as exploited by conjurors) that can run riot where animals are concerned all contribute to a tendency to conclude that animals are cleverer than they really are. Under such circumstances, criticism, and an insistence that simpler explanations must be explored first, are actually constructive rather than destructive steps. But when experiments have been carried out in such a way that alternative simpler 'kill-joy' explanations can be ruled out and we are left with a hard core of studies that make it extremely likely that at least some animals do think in rudimentary ways and that they experience pleasure and suffering sufficiently for these to matter to them, then it seems positively unscientific *not* to consider the possibility that we are looking at the outward and visible signs of inner conscious awareness. Scientific evidence as well as common sense now demand that we take the step of inferring consciousness in species other than our own.

Suppose that we do. What implications does this have for the way we view other species on this planet? There are two ways in which it could be revolutionary. First, depending on where you started out from, it could profoundly change your attitudes to the ways in which animals should be treated. If the beings you eat, hunt, try to eradicate as pests or keep as pets are more 'like us' than you have previously thought, then you may want to revise your beliefs about issues which broadly come under the heading of animal welfare. On the other hand, you may not. I have never seen it as my job to tell people how to behave, merely to provide them with sufficient information that, when they decide for themselves what to do, they base their decision on correct factual information, rather than on an incorrect version of the real world. I am certainly not going to conclude this book with an instruction that nobody should harm another animal or even that they should base their decisions on the extent to which other animals are capable of suffering. The way in which people make moral decisions is often mysterious to me but somewhere in nearly everybody's morality, conscious beings are seen as importantly different from objects with no vestiges of consciousness. So, by providing evidence for complexity of mental processes and the strong possibility of consciousness in non-humans, perhaps I will have made some small difference to the attitudes people take to them, even though I have chosen not to spell out exactly what those attitudes should be. Perhaps all I will have done is to raise a whole new set of questions such as *which* animals we should be concerned about. Is it just chimpanzees and clever parrots that are candidates for consciousness or must all the rest be considered too? Are there lines we can draw somewhere in the vastness of the animal kingdom to separate those that are aware of what they are doing and those that are not?

For what it is worth, I believe these questions do have answers and that those answers will be found in empirical studies of the animals concerned. If it is important to you, in your decisions about how to treat another being, to know how intelligent that being is, then you will need to find out which animals are clever in the sense that matter to your moral code. If, on the other hand, you believe it is more important to know whether a given animal is capable of 'feeling' fear or pain or frustration, then you will need a different set of factual information about it. Either way, the answers will come from biological studies of the animals concerned. Science may not yet know the answers, but it could one day provide them.

For those people who are interested in following up the ethical implications of these discoveries about animal capabilities, I have included at the end a number of books that may be helpful. They take us into the realms of moral decision-making that are beyond the scope of this one, although I hope to have left you, as it were, at the threshold and given you an idea of where you may want to go.

The second implication of acknowledging animal consciousness is that it could completely change what we accept as a valid biological explanation of animals, particularly of their behaviour. If consciousness is a biological phenomenon, evolved because it made animals in some way more effective at getting through their lives, then any explanation that leaves it out must be missing something very important. Many areas of the study of animal behaviour might have to be revised quite radically. If animals are thinking about what they want to communicate to each other, for instance, then maybe the study of animal communication will have to be revised to take this into account. If animals are aware of each other as social individuals with intentions and relationships, then perhaps the study of social organization will have to be modified too. And if decision-making is governed in part by subjective emotions about the most pleasurable or least painful way of proceeding, then this, too, will have to become part of what we accept as an explanation for 'motivation'.

To some extent, these changes have already begun and it may be that they will become much greater as studies of animal behaviour make people change the way they think about beings that just happen not to be human. The study of animal consciousness is already much more accepted than it was even 20 years ago, and we now know a little of what goes on in the minds of animals with the promise of being able to learn a great deal more in the future. I hope that this book has shown that we can proceed on this journey into animal consciousness and still be scientific about it. Doing so is undoubtedly to undertake one of the most exciting enterprises in the whole of biology. Because of the extreme difficulty of the task and the still-mysterious nature of our goal, we must proceed with care and caution. But let us at least proceed.

Further Reading

CHAPTER ONE: Through your eyes only? (pages 1–16)
Animal consciousness.
This issue is discussed by, amongst others:
J.H. Crook (1983) On attributing consciousness to animals. *Nature* **303**, 11–14.
D.R. Griffin (1981) *The Question of Animal Awareness*. New York: Rockefeller University Press.
D.R. Griffin (1992) *Animal Minds*. Chicago: University of Chicago Press.
S. Walker (1983) *Animal Thought*. London: Routledge and Kegan Paul.

Pain
P. Bateson (1991) Assessment of pain in animals. *Animal Behaviour* **42**, 827–40.
H. Rachlin (1985) Pain and behavior. *Behavioral and Brain Sciences* **8**, 43–83.
D.M. Morton and P.H.M. Griffiths (1985) Guidelines on the recognition of pain, distress and discomfort in experimental animals and an hypothesis for assessment. *The Veterinary Record* **116**, 431–6.

CHAPTER TWO: Miss Halsey moves her foot (pages 19–63)

Ostrich egg recognition
B. Bertram (1979) Ostriches recognize their own eggs and discard others. *Nature* **279**, 233–4.

Vervet monkeys
D.L. Cheney and R.M. Seyfarth (1982) How vervet monkeys perceive their grunts. *Animal Behaviour* **30**, 739–51.
D.L. Cheney and R.M. Seyfarth (1990) *How Monkeys See the World*. Chicago: University of Chicago Press.

Red deer stag assessment of each other's fighting ability
T.H. Clutton-Brock and S.D. Albon (1979) The roaring of red deer and the evolution of honest advertisement. *Behaviour* **69**, 145–70.

Appraisal of males by female black grouse
R.V. Alatalo, J. Högland and A. Lundberg (1991) Lekking in the black grouse – a test of male viability. *Nature* **352**, 155–6.

White-throated sparrows recognizing neighbours and strangers
R.J. Brooks and J.B. Falls (1975) Individual recognition by song in white-throated sparrows. I. Discrimination of songs of neighbors and strangers. *Canadian Journal of Zoology* **53**, 879–89.
J.B. Falls and R.J. Brooks (1975) Individual recognition by song in white-throated sparrows. II. Effects of location. *Canadian Journal of Zoology* **53**, 1412–20.

Food-hoarding by birds
D.F. Sherry (1982) Food storage, memory and marsh tits. *Animal Behaviour* **30**, 631–3.
D.F. Sherry (1984) Food storage by black-capped chickadees: memory for the location and content of caches. *Animal Behaviour* **32**, 451–64.
S.J. Shettleworth and J.R. Krebs (1986) Stored and encountered seeds: a comparison of two spatial memory tasks. *Journal of Experimental Psychology: Animal Behavior Processes* **12**, 248–56.

Rats and their avoidance of poisonous foods
N.W. Bond (1984) The poisoned partner effect in rats: some parametric considerations. *Animal Learning and Behavior* **12**, 89–96.
C. Brunton, D.W. Macdonald and A.P. Buckle (in press) Behavioural resistance towards poison baits in brown rats *Rattus norvegicus*. *Applied Animal Behaviour Science*.

B.G. Galef, Jr and M.M. Clark (1971) Social factors in the poison avoidance and feeding behaviour of wild and domesticated rat pups. *Journal of Comparative and Physiological Psychology* **75**, 241–357.

B.G. Galef, Jr (1986) Social identification of toxic diets by Norway rats (*Rattus norvegicus*). *Journal of Comparative and Physiological Psychology* **100**, 331–4.

B.G. Galef, Jr (1991) Information centre of Norway rats: sites for information exchange and information parasitism. *Animal Behaviour* **41**, 295–301.

Decision-making in house sparrows

M.A. Elgar (1986a) The establishment of foraging flocks in house sparrows: risk of predation and daily temperature. *Behavioural Ecology and Sociobiology* **19**, 433–8.

M.A. Elgar (1986b) House sparrows establish foraging flocks by giving chirrup calls if the resources are divisible. *Animal Behaviour* **34**, 169–74.

Food-sharing in vampire bats

G.S. Wilkinson (1984) Reciprocal food-sharing in the vampire bat. *Nature* **308**, 181–4.

CHAPTER THREE: Bees do it (pages 65–102)

The story of Clever Hans

R. Boakes (1984) *From Darwin to Behaviourism: Psychology and the Minds of Animals*. Cambridge: Cambridge University Press.

Zebras and hyenas

H. Kruuk (1972) *The Spotted Hyena: A Study of Predation and Social Behaviour*. Chicago: University of Chicago Press.

Chimpanzees and language

B.T. Gardner and R.A. Gardner (1969) Teaching language to a chimpanzee. *Science* **165**, 664–72.

B.T. Gardner and R.A. Gardner (1975) Early signs of language in child and chimpanzee. *Science* **187**, 752–3.

A. Premack (1976) *Why Chimps Can't Read*. New York: Harper and Row.

D. Premack (1971) Language in a chimpanzee. *Science* 171, 808–22.

D. Premack (1976) *Intelligence in Ape and Man*. Hillsdale, New Jersey: Lawrence Erlbaum.

D. Premack, G. Woodruff and K. Kennel (1978) Paper-marking test for chimpanzee: simple control for social cues. *Science* 202, 903–5.

H.S. Terrace (1979) *Nim*. New York: Alfred A. Knopf.

J. Umiker-Sebeok and T.A. Sebeok (eds) (1980) *Speaking of Apes: A Critical Anthology of Two-way Communication with Man*. New York: Plenum Press.

Bee dances

K. von Frisch (1967) *The Dance Language and Orientation of Bees*. Cambridge, Mass.: Harvard University Press.

M. Lindauer (1971) *Communication among Social Bees*. Cambridge, Mass.: Harvard University Press.

T. Seeley (1977) Measurement of nest cavity volume by the honey beer (*Apis mellifera*). *Behavioural Ecology and Sociobiology* 2, 201–7.

CHAPTER FOUR: Thinking ahead (pages 105–139)

Extrapolation and concepts in pigeons

J.J. Neiworth and M.E. Rilling (1987) A method for studying imagery in animals. *Journal of Experimental Psychology: Animal Behavior Processes* 13, 203–14.

H.S. Terrace (1986) Positive transfer from sequence production to sequence discrimination in a nonverbal organism. *Journal of Experimental Psychology: Animal Behavior Processes* 12, 215–34.

'Counting' in animals

H. Davis and S.A. Bradford (1986) Counting behavior by rats in a simulated natural environment. *Ethology* 73, 265–80.

H. Davis and R. Pérusse (1988) Numerical competence in animals: definitional issues, current evidence and a new research agenda. *Behavioral and Brain Sciences* 11, 561–615.

I. Pepperberg (1987) Evidence for conceptual quantitative abilities in the African grey parrot: labelling of cardinal sets. *Ethology* **75**, 37–61.

Reliability of signals in vervet monkeys
D.L. Cheney and R.M. Seyfarth (1988) Assessment of meaning and the detection of unreliable signals by vervet monkeys. *Animal Behaviour* **36**, 477–86.

Social cleverness in baboons and chimpanzees
R.W. Byrne and A. Whiten (1988) *Machiavellian Intelligence: Social Expertise and the Evolution of Intellect in Monkeys, Apes and Humans*. Oxford: Clarendon Press. The most relevant chapters in this are: Chapter 15: R.W. Byrne and A. Whiten: 'Tactical deception of familiar individuals in baboons', and Chapter 16: A. Whiten and R.W. Byrne 'The manipulation of attention in primate tactical deception'.

H. Kummer (1982) Social knowledge in free-ranging primates. In *Animal Mind–Human Mind* (ed. D.R. Griffin) 113–20. Berlin, Heidelberg and New York: Springer-Verlag.

E.W. Menzel (1974) A Group of young chimpanzees in a 1-acre field: leadership and communication. In *Behavior of Non-human Primates* (ed. A.M. Schrier and F. Stollnitz, Vol. 5, pp. 83–153. New York: Academic Press. (Reprinted in *Machiavellian Intelligence*, pp. 155–9.)

CHAPTER FIVE: Feeling our Way (pages 141–164)

What animals work to gain or avoid
G.S. Losey and L. Margules (1974) Cleaning symbiosis provides a positive reinforcer for fish. *Science* **184**, 179–80.

A.P. Silverman (1978) Rodents' defence against cigarette smoke. *Animal Behaviour* **26**, 1279–81.

Preferences of battery hens
F.W.R. Brambell (1965) Chairman. Report of the Technical Committee to Enquire into the Welfare of Animals kept under Intensive Livestock Husbandry Systems. Cmnd 2836. London: HMSO.

N. Bubier (1990) Behavioural priorities in laying hens. D.Phil. thesis, University of Oxford.

B.O. Hughes and A.J. Black (1973) The preference of domestic hens for different types of battery cage floor. *British Poultry Science* 14, 615–19.

Working pigs
L. Matthews and J. Ladewig (1987) Stimulus requirements of housed pigs assessed by behavioural demand functions. *Applied Animal Behaviour Science* 17, 369 (abstract).

Fighting fish
J.A. Hogan, S. Kleist and C.S.L. Hutchings (1970) Display and food as reinforcers in the Siamese fighting fish (*Betta splendens*). *Journal of Comparative and Physiological Psychology* 70, 351–7.

Emotions, feelings and behaviour
M. Cabanac (1979) Sensory pleasure. *Quarterly Review of Biology* 54, 1–29.

M. Cabanac and K.G. Johnson (1983) Analysis of a conflict between palatability and cold exposure in rats. *Physiology and Behavior* 31, 249–53.

K. Oatley (1989) The importance of being emotional. *New Scientist*, 19 August, 33–6.

CHAPTER SIX: Balance of evidence (pages 167–181)

Some discussions of consciousness
J. Baars (1988) *A Cognitive Theory of Consciousness*. New York and Cambridge: Cambridge University Press.

P.S. Churchland (1986) *Neurophilosophy: Towards a Unified Science of the Mind/Brain*. Cambridge, Mass.: MIT Press.

D. Dennett (1991) *Consciousness Explained*. Boston: Little, Brown.

D.R. Griffin (1976) *The Question of Animal Awareness*. New York: Rockefeller University Press.

N. Humphrey (1983) **Consciousness Regained**. Oxford: Oxford University Press.

M. Lockwood (1989) *Mind, Brain and the Quantum: The Compound Eye*. Oxford: Basil Blackwell.

A.J. Marcel and E. Bisiach (eds) (1988) *Consciousness in Contemporary Science*. Oxford: Clarendon Press. This book contains several useful essays, including: A. Allport 'What concept of consciousness?' (pp. 159–182); K. Oatley 'On changing one's mind: a possible function of consciousness' (pp. 369–389); and P.N. Johnson-Laird 'A computational analysis of consciousness' (pp. 357–368).

T. Natsoulas (1978) Consciousness. *American Psychologist* 33, 906–14.

R. Penrose (1989) *The Emperor's New Mind*. Oxford: Oxford University Press.

Implications for animal welfare

M. Bekoff and D. Jamieson (1991) Reflective ethology, applied philosophy, and the moral status of animals. In *Perspectives in Ethology* Vol. 9 (eds P.P.G. Bateson and P.H. Klopfer), pp. 1–47. New York and London: Plenum Press.

J.W. Driscoll and P. Bateson (1988) Animals in behavioural research. *Animal Behaviour* 36, 1569–74.

G.G. Gallup and J.W. Beckstead (1988) Attitudes toward animal research. *American Psychologist* 43, 474–6.

M. Midgley (1983) *Animals and Why They Matter*. London: Penguin Books.

T. Regan (1983) *The Case for Animal Rights*. Berkeley: University of Californian Press.

P. Singer (1975; revised 1990) *Animal Liberation: A New Ethics for our Treatment of Animals*. New York: Avon.

Behaviourism and its grip on studies of animal behaviour
This is well described by:

R. Boakes (1984) *From Darwin to Behaviourism: Psychology and the Minds of Animals*. Cambridge: Cambridge University Press.

H. Gardner (1985) *The Mind's New Science: A History of the Cognitive Revolution*. New York: Basic Books.

Acknowledgements for Illustrations

FRONTISPIECE	*The Telegraph Colour Library*
CHAPTER ONE (ALLIGATOR)	*Mike Amphlett*
CHAPTER TWO (OSTRICH)	*Mike Amphlett*
CHAPTER THREE (ZEBRAS)	*The Telegraph Colour Library*
CHAPTER FOUR (HERON)	*Mike Amphlett*
CHAPTER FIVE (RABBIT)	*Mike Amphlett*
CHAPTER SIX (RACOON)	*The Telegraph Colour Library*

Index